Advanced Techniques for Anomaly Detection

This book is a comprehensive guide that explores the latest developments in anomaly detection techniques across a range of fields, including cybersecurity, finance, image processing, sensor networks, social network analysis, health systems, and IoT systems. With 6 chapters covering various topics such as deep learning-based anomaly detection, feature selection and extraction techniques, ensemble methods, and evaluation metrics, this book offers a comprehensive understanding of advanced anomaly detection techniques and their applications in different fields. This book will be an excellent resource for researchers, practitioners, and students interested in anomaly detection and its applications in various domains.

Abdul Wahid, PhD, Department of Computer Science and Engineering, Indian Institute of Information Technology Dharwad, India.

Praveen Kumar Donta, PhD, Department of Computer and Systems Sciences, Stockholm University, Sweden.

Advanced Techniques for Anomaly Detection

Beyond the Basics

Edited by
Abdul Wahid and Praveen Kumar Donta

CRC Press
Taylor & Francis Group
Boca Raton London New York

CRC Press is an imprint of the
Taylor & Francis Group, an **informa** business

Front cover image: petrov-k/Shutterstock

First edition published 2025
by CRC Press
2385 NW Executive Center Drive, Suite 320, Boca Raton FL 33431

and by CRC Press
4 Park Square, Milton Park, Abingdon, Oxon, OX14 4RN

CRC Press is an imprint of Taylor & Francis Group, LLC

ISBN: 978-1-032-72930-5 (hbk)
ISBN: 978-1-032-73298-5 (pbk)
ISBN: 978-1-003-46355-9 (ebk)

DOI: 10.1201/9781003463559

Typeset in Minion
by Deanta Global Publishing Services, Chennai, India

Contents

CHAPTER 1

Anomaly Detection in Carbon Sequestration

Advancements, Challenges, and Future Directions

M. Saif, Raj Kiran, Prabhat Dansena, and Prasun Chandra Tripathi

1.1 INTRODUCTION

As per the Paris Agreement 2015, the ultimate objective is to limit the increase in the average global temperature to below 2°C and to make an effort to limit the temperature increase to 1.5°C above pre-industrial levels. There are multiple existing pathways to achieve these objectives. One approach includes restricting CO2 emissions by capturing them from the atmosphere and sequestering them. Soil, plants, and oceans are natural intervening agents controlling the level of CO2, while artificial interventions can be designed to capture and subsequently utilise/store it in some form of product. Artificial interventions include the utilisation of CO2 in various industries such as the production of chemical products (formic acid, green cement), firefighting, and water treatment. In this sequence, geological carbon sequestration is another avenue to deal with climate change challenges and accomplish the above-mentioned.

This chapter discusses the techniques and implementation of machine learning for geological carbon sequestration (GCS). Geological carbon sequestration refers to the process of injecting captured CO2 into geological formations such as depleted oil and gas reservoirs, producing oil and gas reservoirs, coal seams, saline aquifers, and mafic or ultramafic rocks

DOI: 10.1201/9781003463559-1

(Saif, Kiran, Rajak, & Verma, 2024). In such processes, the subsurface-to-surface migration of stored gas is a unique safety challenge pertaining to the containment security in the reservoir. Such challenges can be addressed by monitoring and mitigating any leaks, losses, or disturbances that can allow re-entrainment of previously stored CO2 back into the atmosphere. As per the increasing focus on such mechanisms to address climate challenges, well-organised safety and monitoring protocols need to be established to safeguard the subsurface integrity of injected CO2. This involves monitoring the surface emissions in the soil, forests, and CCS facilities to detect any signatures of gas emissions in the form of anomalies (irregularities from normal behaviour) and, consequently, act according to such unforeseen circumstances.

The main objective of this chapter is to deliberate upon anomaly detection techniques specifically in the context of geological carbon sequestration. Anomaly detection can be an essential component of assuring the dependability and success of carbon sequestration projects. Using data, technology, and monitoring systems, abnormal or troublesome changes in the spatial-temporal behaviour of CO2 migrating to the surface can be identified. Through an investigation of the fundamentals, techniques, and practical uses of anomaly detection, this chapter presents the tools and methods which will support carbon sequestration activities, eventually bolstering the objectives of climate change mitigation.

1.2 GEOLOGICAL CARBON SEQUESTRATION

Carbon capture and sequestration (CCS) facilities are man-made systems that store the captured CO2 emissions and consequently stop the emissions from power stations or industrial activities from escaping into space. Studies suggest that approximately 9.5 gigatons (Gt) of carbon/year, or 35 Gt CO2/year, must be permanently sequestered to achieve the future goal of keeping global warming below 2°C over pre-industrial levels (Le Quéré et al., 2015). The usual steps in the process of CCS include the capture of CO2 from either source or from the atmosphere, transportation to the storage sites, and injection into the subsurface. Geological carbon sequestration (GCS) refers to the subsurface CO2 storage process, which is conducive for greenhouse gas emissions due to its large storage capacity and stability over a long period of time. While there are many advantages to this approach, there are also some specific problems like the possibility of injection well leakage, geological subsurface movement (tectonic effect), and reservoir pressure control, and therefore it becomes essential to monitor and identify the anomalies to preserve the security and efficiency of GCS.

Consistent efforts have been put forth to establish secure and reliable CO2 sequestration technology in deep geologic formations to mitigate the greenhouse gas emissions in the atmosphere. There are several target zones for geologic sequestration, including but not limited to unmineable coal seams, saline formations, and depleted hydrocarbon reservoirs (Car). Leakage via poorly sealed and abandoned wells is thought to be the most likely route for CO2 migration (Gasda, Bachu, & Celia, 2004). Massive-scale subsurface pressurisation could result from the injection of large volumes of CO2, and if brine or injected fluids migrate unintentionally, it could have a negative impact on the natural resources nearby (Birkholzer & Zhou, 2009; Pruess, 2004). In recent years, numerous risk management models and techniques have been put forth to monitor and mitigate such conditions. These models and techniques include scenario-based methods and life-cycle management-based strategies. The scenario-based methods incorporate key system elements, events, and processes to determine the probabilities and outcomes using stochastic simulation or expert elicitation (Kopp, Binning, Johannsen, Helmig, & Class, 2010; Oldenburg, Bryant, & Nicot, 2009; Stauffer, Viswanathan, Pawar, & Guthrie, 2009). The life-cycle management-based strategy uses monitoring, verification, and accounting to evaluate the migration parameters of CO2 plumes at various operational stages and ensures the containment in storage formations (Car, dnv, Benson, 2007). In these techniques, monitoring plays a pivotal role in the GCS projects to reduce the possibility of unintentional CO2 leakage to the surface from storage formations (Chen, Harp, Lin, Keating, & Pawar, 2018; Haszeldine, 2009; Jenkins, Chadwick, & Hovorka, 2015; Zhong & Carr, 2019).

Additionally, other tools and methods such as geochemical and geophysical surveys, pressure and temperature monitoring, soil gas isotopic analysis, groundwater monitoring, remote-sensing, and tracers have been developed to monitor and mitigate the possibility of leakage and gas migration (Lemieux, 2011; Lewicki, Hilley, & Oldenburg, 2005; Romanak, Bennett, Yang, & Hovorka, 2011; Rutqvist, Vasco, & Myer, 2010; Seto & McRae, 2011). Using the underlying physical processes and real-time data, numerical and analytical models have been constructed for leakage risk assessment and deployment of suitable response strategy for such migration scenarios (Celia, Nordbotten, Court, Dobossy, & Bachu, 2011; Cihan, Zhou, & Birkholzer, 2011; Nordbotten, Celia, Bachu, & Dahle, 2005b; Oldenburg & Unger, 2003; Zhou, Birkholzer, Tsang, & Rutqvist, 2008). Consequently, these models are further used to identify the uncertainty in in situ and operational parameters (Buscheck et al., 2012; Oladyshkin, Class, Helmig, & Nowak, 2011; Sun, Zeidouni, Nicot, Lu, & Zhang, 2013; Sun & Nicot, 2012).

Leakage monitoring is crucial for the success of the GCS project. Various leakage pathways can exist in the subsurface environment due to the establishment of hydraulic connectivity from the storage zones. This hydraulic connectivity can be due to the presence of near-by faults, interzonal connections, and improper sealing of wellbore elements. The pressure monitoring becomes the most reliable tool to identify the spatial-temporal volumetric leakage (Jung et al., 2013; Sun & Nicot, 2012). These downhole pressure monitoring tools are deployed at above-zone monitoring intervals (AZMI), which aids in the early leak detection (Hovorka, Meckel, & Trevino, 2013). The proximity of AZMI to the target zone can transmit the pressure disturbances quickly (Nordbotten, Celia, & Bachu, 2004). The pressure disturbances can be correlated with the leakage signals depending on the monitoring techniques.

The processes/parameters pertaining to the leakage risks are identified during the assessment, which can help in identifying the possibility of CO2 plume or pressure build-up region reaching the wells (Jordan, Oldenburg, & Nicot, 2011; Kopp et al., 2010; Oladyshkin et al., 2011; Oldenburg et al., 2009). Various mathematical models and solutions have been developed over the decades to assess the consequential pressure buildup, leakage rates, and potential damages due to CO2 injections (Ebigbo, Class, & Helmig, 2007; Humez, Audigane, Lions, Chiaberge, & Bellenfant, 2011; Nordbotten et al., 2004; Nordbotten, Celia, & Bachu, 2005; Nordbotten, Celia, Bachu, & Dahle, 2005; Pruess, 2004, 2008). Nordbotten et al. (2004) presented an analytical model to forecast the amount of leakage flux through the abandoned wells (Nordbotten et al., 2004). Vishwanathan et al. (2008) developed the wellbore leakage model (Viswanathan et al., 2008). Zeidouni (2012) and Wang, Bai, Li, Liu, Wu, and Hu (2016) presented the analysis on the possibility of leakage along pre-existing faults (Zeidouni, 2012; Wang, Bai, Li, Liu, Wu, & Hu, 2016). Sun, Lu, and Hovorka (2015; Sun, Lu, Freifeld, Hovorka, & Islam,2016; Sun, Lu, & Islam, 2017) proposed and implemented field-scale pressure-based pulse testing methodology to probe leakage channels in storage formations (Sun, Lu, & Hovorka, 2015, Sun, Lu, Freifeld, Hovorka, & Islam, 2016; Sun, Lu, & Islam, 2017). Yang, Dilmore, Bromhal, and Small (2018) developed risk-based leakage monitoring adaptive design methodology to capture the probabilities of such events (Yang, Dilmore, Bromhal, & Small, 2018). They further implemented a decision-making tree approach to predict the likelihood of leakage occurrence.

Lewicki, Hilley, and Oldenburg (2005) implemented a background noise filtering technique along with near-surface CO2 flux/concentration measurement to model spatial-temporal behaviour of surface CO2 leakage. The filtering technique included the Gaussian weighting function-based spatial coherence and temporal averaging to detect the location and magnitude of surface CO2 leakage (Lewicki, Hilley, & Oldenburg, 2005). Fessenden, Clegg, Rahn, Humphries, and Baldridge (2010) presented an overview and findings of three novel systems to detect and identify surface seepage during a controlled CO2 release experiment at Bozeman facility. These three systems include an in situ frequency modulation spectroscopy system, O2/CO2 analysis system, and long-range alpha detection (LRAD) system. However, these systems require identification of signatures corresponding to associated leakage and disturbances (Fessenden, Clegg, Rahn, Humphries, & Baldridge, 2010). Therefore, it becomes important to understand the anomaly detection from the associated data.

1.2.1 Anomaly Detection for Carbon Sequestration

The success of carbon sequestration projects depends on the quick identification of problems and emergencies that may hinder their efficiency and reliability. Monitoring and verification systems are progressively becoming more and more essential for carbon sequestration projects which rely on data-driven modelling. Data-driven modelling incorporates anomaly detection, which helps in mitigating any possible hazards such as leakage and geological shifts due to tectonics. This involves finding patterns or data points in a dataset that significantly differ from the expected or typical behaviour. It helps identify abnormal or unexpected occurrences concerning preservation and storage of carbon dioxide in sequestered form. The detection of anomalies becomes important in carbon sequestration for several reasons. It can help protect the integrity of carbon sequestration operations by spotting problems and abnormalities as early as possible. Interventions may be designed on such signatures to establish long-term reliance in such cases. Furthermore, it can also help in preventing unintentional releases of stored CO2 due to operational inefficacy. Remedial works may be possible when anomalies indicate problems such as soil disturbances, deforestation, or leakage in CCS facilities.

Robust anomaly detection is driven by several factors including data sources, technology and tools, and algorithms. Anomaly identification depends on many sources of data including temperature and pressure readings, and CO2 concentration at the storage sites in real time. Geophysical surveys and geological monitoring data can provide subsurface information about in situ downhole conditions. Remote sensing data will help us keep an eye on geological storage locations so that any abnormalities can be identified at the surface. Anomaly detection systems should incorporate fail-safe advanced tools and technologies. Large datasets can be processed to identify abnormalities or deviations from normal behaviour. Several anomaly detection methods and algorithms such as time-series analysis, statistical techniques, and machine learning models like artificial neural networks (ANN) and support vector machines (SVM) can be used to process these datasets. These models provide insight into trends and patterns within the data that help with anomaly identification ultimately.

1.2.1.1 Basic Principles of Anomaly Detection Techniques

Different strategies are used by anomaly detection algorithms to detect anomalies in datasets including baseline establishment, feature selection, algorithm selection, threshold setting, continuous monitoring, and response and action, as shown in Figure 1.1. The details of these strategies are:

(1) Establishing a baseline: The first step involves identifying 'normal' behaviour for a dataset, which is usually achieved by using historical data analysis. Based on legacy operational data, a baseline of expected values or patterns can be constructed.

(2) Selection of features: Anomaly can assume multiple forms, and thus it is important to select and investigate the relevant characteristics (properties) of the data. For example, attributes characteristic of geological carbon sequestration may include features such as porosity, permeability, rock type, sealing effectiveness, and depth along with pressure and temperature conditions.

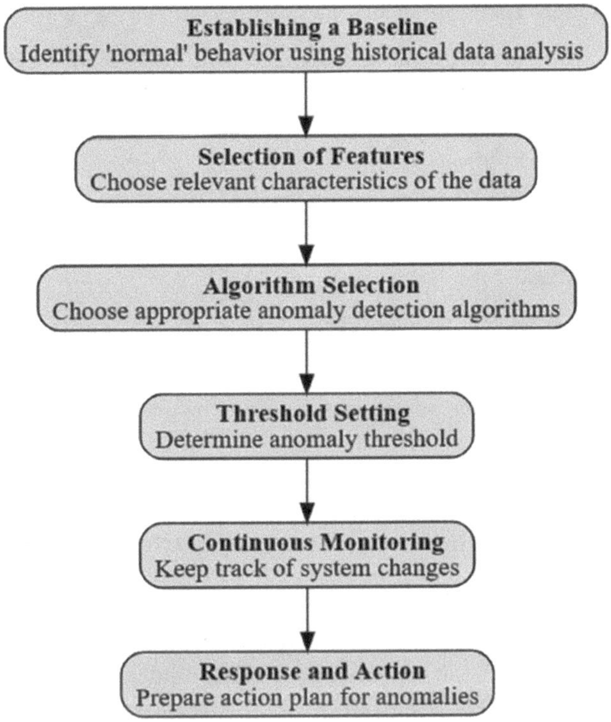

FIGURE 1.1 Flow chart for the anomaly detection method.

(3) Algorithm selection: Time-series analysis, machine learning models (Isolation Forests and One-Class SVM), or statistical techniques (Z-scores) are a few algorithms available for anomaly detection. Selection depends on the nature of data and types of anomalies.

(4) Threshold setting: In anomaly detection methods, an often-employed step is determining a threshold value at which certain number(s) of observations are considered anomalous. This threshold could be set based on statistical significance or domain knowledge.

(5) Continuous monitoring: The monitoring of the real-time data will help in identifying the abnormal changes in the system.

(6) Response and action: Appropriate response and action plan will be required in case of any anomaly within the system. This requires the evaluation of various leakage scenarios.

1.2.1.2 Anomaly Detection in the CCS Project

Anomaly detection systems are important components of CCS facilities that enable safe and efficient CO2 sequestration. They can be implemented using the following steps:

(1) Data monitoring: These systems track CO2 concentrations, flow rates, temperature, and pressure through several sensors and devices at the CCS facilities.

(2) Baseline establishment: Similar to regular anomaly detection system, the baseline needs to be established with historical data defining normal operating conditions.

(3) Real-time analysis: The system incorporates real-time data analysis by comparing the received data with a pre-established baseline. An alert can be generated whenever there is any departure from the anticipated values or patterns.

(4) Alert generation: The system notifies the facility operators via an alert or notice if an anomaly is found. The nature and location of the abnormality can be disclosed in such alert or notice.

(5) Response automation: Anomaly detection systems can be accompanied with automated reactions including stoppage of specific activities, turning on safety precautions, or starting remedial measures.

1.2.1.3 Importance of Early Detection of Leaks and Malfunctions

For multiple reasons, it is critical to identify leaks and failures in CCS facilities as soon as possible (as shown in Figure 1.2). Some reasons are as follows:

(1) Safety: Leaks of CO2 can be harmful to people's health, and therefore early leak and malfunction detection protects the people working at the facility, the neighbourhood, and the environment.

(2) Environmental protection: The unintentional release of stored CO2, a greenhouse gas, can be avoided by prompt detection. The environmental advantages of CCS could be offset by a significant leak.

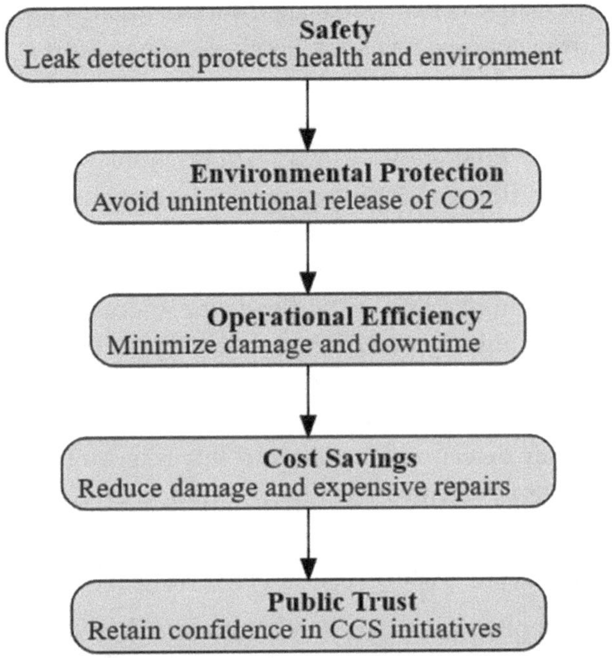

FIGURE 1.2 Importance of early leakage detection in CCS project.

(3) Operational efficiency: Early problem detection minimises potential equipment damage, reduces downtime, and enables timely repair action.

(4) Cost savings: Quick action in the event of an anomaly can reduce damage and save expensive repairs or environmental cleanup.

(5) Public trust: Retaining public confidence is critical to the acceptability and effectiveness of CCS initiatives. It shows ethical and safe carbon storage in promptly addressing abnormalities.

1.3 DATA AND TECHNOLOGY

Numerous researchers have examined and applied neural networks for anomaly identification. Major computer-based anomaly detection techniques were compared and studied, including fuzzy logic-based and Bayesian systems, neural networks, and genetic algorithms. One of the main benefits of neural network-like algorithms is their capacity to generalise patterns from sparse data (Kaur, Singh, & Minhas, 2013). In the CCS project, the pressure data is frequently non-stationary, multi-temporal, and stochastic because of variations in field operations, such as surface pump rates. However, the pressure waveform exhibits fundamental predictable patterns, such as rising and falling pressures in response to increases in injection rates and shut-in period, respectively. Therefore, one must use squared error as a metric and neural network in conjunction with a sliding window. Using inputs like traffic flows, the anomalous traffic can be identified by using multilayer feedforward neural network (MFNN) architecture, and the outcomes should be compared with a threshold (Callegari, Giordano, & Pagano, 2014). Using long short-term memory (LSTM) networks, the anomaly detection for unknown intrusion can also detected in computer networks (Bontemps, Cao, McDermott, & Le-Khac, 2016).

1.3.1 Multilayer Feedforward Neural Network (MFNN)

To construct a complex multi-layer model, modern deep learning architectures use numerous layers of neurons. A layer is made up of an artificial neural network (ANN), and MFNN is one such network architecture. The network needs an input layer, an output layer, and one or more hidden layers. The layer that takes in input is known as the input layer, the layer that generates the intended prediction is known as the output layer, and the hidden layer is an intermediate layer that captures complex patterns in the data. The loss function (also known as cost/objective function) in MFNN is used to quantify the difference between the network's prediction and actual target values. Loss functions (e.g., mean squared error (MSE), cross-entropy loss, binary cross-entropy loss, and hinge loss) provide a measure of how well the network is performing with the goal of minimising the loss during training. Based on a loss function measure, the network learns via the backpropagation method (Mozer, 2013; Sathyanarayana, 2014). A typical MFNN system is presented in Figure 1.3.

MFNN is a type of feedforward ANN that utilises neurons as its fundamental computing unit (Amid & Mesri Gundoshmian, 2017; Rynkiewicz, 2012; Yilmaz & Kaynar, 2011). An artificial neuron is a processing unit made up of weighted inputs and outputs. An activation function adds more modification to the output signal. In simple linear regression, the weights on the inputs are comparable to the regression coefficients. These are the ideal

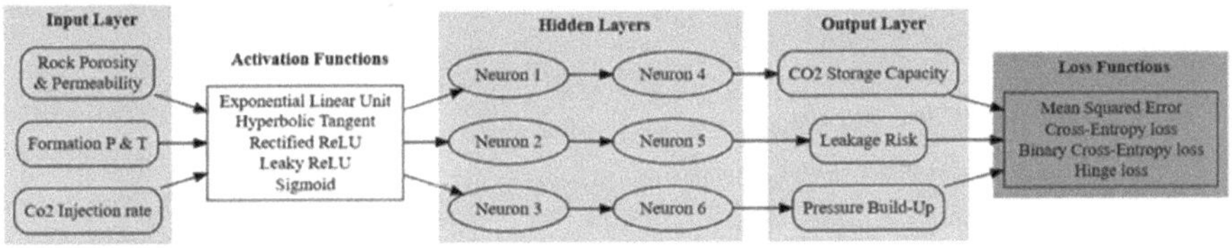

FIGURE 1.3 Model architecture for MFNN in GCS.

model parameters, and they are frequently started with random values and adjusted as the model learns. In ANN models, the non-linear component is provided via an activation function. Over the years, a variety of activation functions have been developed, for e.g., sigmoid, hyperbolic tangent, rectified linear unit (ReLU), leaky ReLU, and exponential linear unit (ELU).

1.3.2 Long Short-Term Memory (LSTM)

One type of recurrent neural network (RNN) is the LSTM neural network, as shown in Figure 1.4. RNN features an internal feedback loop that can retain memory over longer data sequences (Hochreiter et al., 2001; LeCun et al., 1995; Nguyen et al., 2020). Time-series prediction is thus a natural fit for the RNN family of network designs. Rather than a single neuron, the network is made up of memory blocks that connect to form dense layers by sequencing. The memory block is gated, which means that each gate has a unique way of activation. Usually there are three gates that can be found in a block: an input gate, an output gate, and a forget gate. The input gate modifies the memory state, the output gate determines the output of memory block, and the forget gate takes decision on which information to keep or discard. The collection of gated memory blocks and cells allows the network to read data sequences rather than simply a single point.

Originally, LSTM was designed to address the vanishing/exploding gradient issue, which causes RNN training divergence (Hochreiter & Schmidhuber, 1997). Similar to RNN, LSTM is very good at capturing dynamic properties in the graph through cycles (Ma & Hovy, 2016). Additionally, in LSTM, the number of unknowns are reduced significantly due to usage of the same parameters (network weights) at every time step (Che, Purushotham, Cho, Sontag, & Liu, 2018).

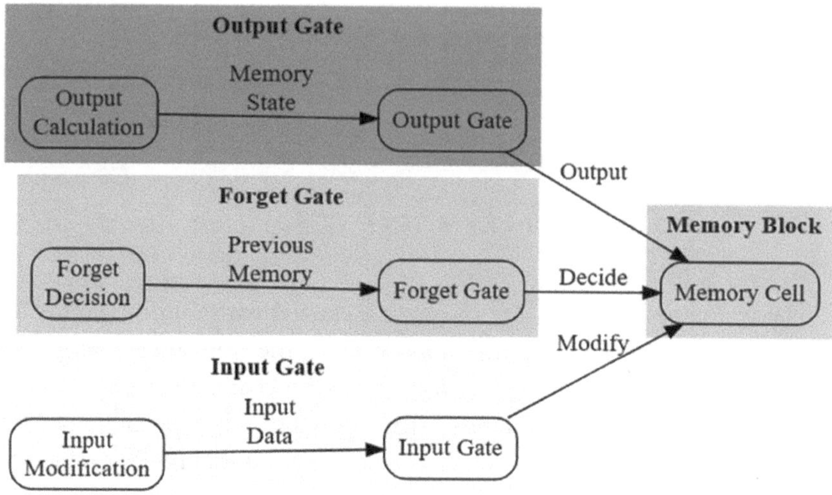

FIGURE 1.4 LSTM architecture.

1.3.3 Convolutional Neural Networks (CNN)

One of the neural network layers in deep learning models that is most frequently employed is CNN. The capacity of CNN to effectively extract characteristics from an input image through convolution computations to create so-called feature maps is its greatest strength. Three fundamental constructs: weight sharing, downsampling, and local receptive fields, make up a typical CNN building block. It is not feasible to connect every neuron in a layer to every other neuron in its input layer when working with high-dimensional imagery. Each neuron in the preceding layer can only connect to a limited neighbourhood due to its local receptive field. Each neuron may extract local information from the input image, like line segments, end-points, and corners, by using these local receptive fields. The typical CNN architecture is provided in Figure 1.5. One technique to drastically lower the number of parameters that need to be optimised during network training is through weight sharing. To obtain features from an input image, a moving window (also known as a filter or kernel) essentially uses the same set of weights, based on the idea that a feature which is helpful in one place would likely be helpful in other places as well.

In other words, the three sets of layers which often found in a CNN architecture are convolutional, pooling, and fully linked layers. To create

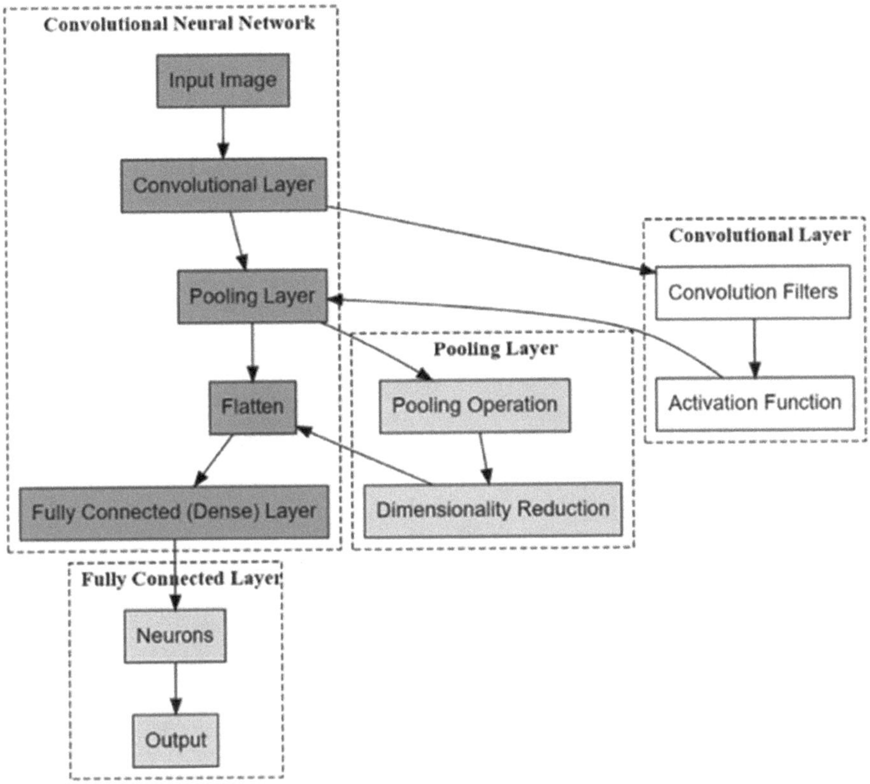

FIGURE 1.5 LSTM architecture.

a filtered output, a convolutional layer convolves the input through several filters. These filters aid in the extraction of intricate aspects from the data, such as the pressure's sinusoidal pattern. After that, the filtered result is frequently fed into a pooling layer to lower the dimensionality of the data. After being flattened, the pooling layer's output is sent to a fully connected dense layer or set of layers (Gu et al., 2018, Khan, Sohail, Zahoora, & Qureshi, 2020; Simonyan & Zisserman, 2014; Zeiler & Fergus, 2014). To enforce translation invariance, an operation called down-sampling or pooling is used. This means that the same feature stays active even when the image undergoes slight translations. Unlike MFNN, only a tiny portion of the output will be connected by the neurons in a CNN architecture from the preceding neurons, i.e., just from a chosen patch from the previous layer, thus decreasing the likelihood of overfitting. The strong ability of CNN to

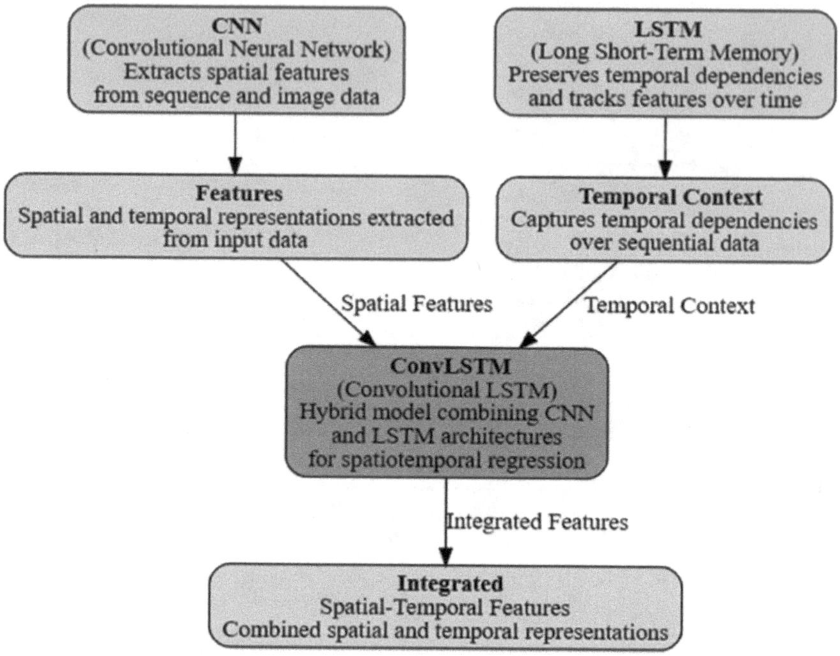

FIGURE 1.6 Convolutional LSTM model outline.

extract high-level features from spatial space makes it useful to extract the spatial information at each gate and cell states.

1.3.4 Convolutional LSTM (ConvLSTM)

ConvLSTM, a hybrid CNN and LSTM model, combines the best features of both models to provide intense representations of complex spatiotemporal processes, as shown in Figure 1.6. These representations can then be used as a deep learning-based proxy model to reflect the underlying physics. ConvLSTM has been used to detect human movements (Ji, Xu, Yang, & Yu, 2012), for rainfall nowcasting (Shi, Chen, Wang, Yeung, Wong, & Woo, 2015), for visual recognition and description (Donahue et al., 2015), and to recognise body gestures (Zhu, Zhang, Shen, & Song, 2017). Few researchers have implemented ConvLSTM to identify the subsurface anomaly detection issues thus far.

LSTM is beneficial because it preserves the temporal aspect of sequence data. However, CNN's design is strong enough to extract features from sequence and picture data alike. As a result, for the pressure predictions, we ought to use a hybrid CNN/LSTM model (ConvLSTM). To extract complicated forms from the time-series data, CNN was used, and LSTM was used to follow these features over longer periods. Hybrid ConvLSTM models are highly effective at completing both tasks (Shi, Chen, Wang, Yeung, Wong, & Woo, 2015). The inability of LSTM to include spatial information is a significant limitation. ConvLSTM uses a CNN and LSTM combination to address this issue. All of the LSTM's inputs are considered as tensors, cell outputs, hidden states, and gates in ConvLSTM. Reservoir variables which are both static and dynamic may be denoted using the high-dimensional arrays known as tensors.

1.4 CHALLENGES AND FUTURE DIRECTIONS

There are a several difficulties and restrictions associated with anomaly identification in the carbon sequestration project, especially in geological formations. The following are few limitations associated with anomaly detection modelling.

1.4.1 Limitations

(1) Data acquisition and monitoring: Zhong et al. (2019) noted that real-time monitoring of the carbon storage reservoirs is essential in order to determine if CO2 leaks and to guarantee success in carbon sequestration projects. Nevertheless, it may be difficult to train and evaluate anomaly detection models due to low amounts and quality of collected data from the reservoirs (Zhong, Sun, Yang, & Ouyang, 2019).

(2) Complexity of geological formations: Song, Jun, Na, Kim, Jang, and Wang (2023) confirmed that GCS sites are often complicated systems where numerous variables control CO2 storage, temperature, pressure, and several geological characteristics. Hence, due to the complexity of this system, anomalies in it can be hard to identify and predict (Song, Jun, Na, Kim, Jang, & Wang, 2023).

(3) High costs: As per Government Accountability Office, US, CCUS technologies are still at their infancy stage and have several impediments towards their wide acceptance, including the high cost. This

could curtail the progress and use of advanced complex anomaly detection methods for carbon sequestration initiatives.

(4) Expertise and collaboration: Kelemen, Benson, Pilorgé, Psarras, and Wilcox (2019) state that multiple disciplines—oil and gas industry professionals, academic institutions officials, and government agencies representatives, among others—must come together in order to detect abnormality in carbon sequestration projects. However, due to different goals, resources, and levels of experience, this collaboration may also provide complications (Kelemen, Benson, Pilorgé, Psarras, & Wilcox, 2019).

(5) Leakage detection: In CCS projects, anomaly detection is often focused on finding the sequestered CO2 leakage from the reservoir (Sinha et al., 2020). Leakage incidents, however, can be hard to find, particularly in the early phases, and we may need advanced monitoring and analysis methods.

Although there are obstacles, technological developments and stakeholder cooperation can help in overcoming some of the constraints by, for instance, making use of pressure readings from the base experiment with deep learning technology, which has been developed for anomaly detection in GCS locations (Gulf Coast Carbon Center, U. of A., 2020). Additionally, pressure field data has been used to apply machine learning approaches to detect CO2 leakage in CCS projects (Sinha et al., 2020), and hence anomaly detection can be made more successful with additional research and development in these areas.

1.4.2 Emerging Trends

Future projections and evolving trends in anomaly detection and CCS projects are driven by the geomechanical problems related to geological CO2 sequestration, the status of current technology, and the improvement of machine learning techniques.

(1) Machine learning approaches: Hussin, Rahim, Hatta, Aroua, and Mazari (2023) performed an in-depth evaluation of the machine learning techniques in applications of carbon capture, which highlights the increasing curiosity in using machine learning for several CCS applications such as anomaly detection. This trend specifies

that using advanced machine learning approaches can enhance monitoring and risk assessment, which can progressively influence the anomaly detection in carbon sequestration projects.

(2) Status of available technologies: Future projections for anomaly detection in carbon sequestration are appreciably influenced by the state of existing carbon capture systems and the potentials for utilising or storing captured CO2. The advancement and execution of sophisticated anomaly detection techniques will undoubtedly be influenced by these technologies. The readiness for wider demonstration or deployment, as well as continued efforts of reducing high costs, can be solved easily.

(3) Geomechanical challenges: A proper comprehension of the geomechanical challenges involved in GCS projects is necessary to assess their feasibility and safety (Song, Jun, Na, Kim, Jang, & Wang, 2023). The ability to address these issues will define the prospects for the industry, which could affect the pace of advancement and implementation of GCS-specific anomaly detection techniques.

In summary, carbon capture technology status, overcoming geomechanical barriers, and rising trends plus opportunities in anomaly detection for carbon sequestration are interlinked. In the CCS context, these traits are expected to result in more efficient and reliable anomaly detection and threat systems (ANDTs) that detect anomalies much faster.

1.4.3 Potential Improvements

New research suggests potential innovations and improvements in CCS systems anomaly detection method, particularly regarding machine learning and deep learning (Figure 1.7). Some of these improvements include but not limited to are:

(1) Anomaly detection as leakage detection: Leakages in the carbon sequestration projects are considered an anomaly detection problem wherein deviations from an expected or normal behaviour indicate reservoir leakage (Sinha et al., 2020). Anomaly detection techniques of machine learning can be employed to improve and enhance the detection of anomalies that are associated with CO2 leaching from reservoirs.

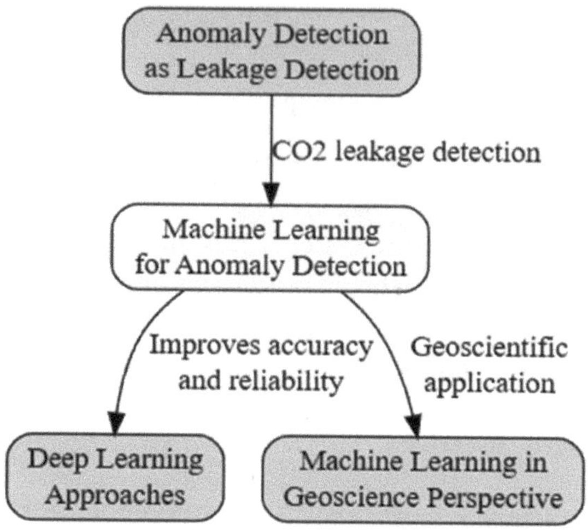

FIGURE 1.7 Implications of leakage detection in carbon sequestration.

(2) Deep learning approaches: Pressure measurements at GCS sites have been combined with deep learning methods that suggest that advanced computing can be creatively applied in this context (Zhong, Sun, Yang, & Ouyang, 2019). Use of deep learning approaches improves the accuracy and reliability of anomaly detection systems for GCS projects.

(3) Machine learning from the geoscience perspective: A broad review of the geoscientific application of machine learning in CCS highlights the potential for multidisciplinary approaches to improve anomaly identifications (Yao, Yu, Zhang, & Xu, 2023). Hence, more efficient anomaly detection solutions to the unique problems of GCS sites may result from combining the geoscientific knowledge with machine learning approaches.

With these understandings, possible advancements and improvements to enhance the anomaly detection method could involve integrating geoscientific knowledge to create more reliable and precise anomaly detection systems for carbon sequestration reservoirs, in addition to investigating the techniques of machine learning and deep learning further. These advancements can support continued efforts to guarantee the efficiency and safety of carbon sequestration initiatives.

1.5 CASE STUDIES

A collection of case studies offers actual viewpoints on the critical role anomaly detection has played in various GCS initiatives. These case studies highlight the importance of anomaly detection in maintaining the integrity of carbon storage and show how it can be used practically in GCS projects. Some of the GCS projects that have been successfully implemented worldwide are discussed below.

1.5.1 Sleipner CO2 Storage Project (Norway)

The Sleipner project is located in the North Sea and incorporates injection of CO2 in saline aquifer. The project started in 1996 and stored about 10 Mt of CO2 by mid-2008 in saline aquifer safely (Shukla, Ranjith, Haque, & Choi, 2010). The project incorporated a variety of techniques including seismic measures to ensure confinement of CO2 into the reservoir (Solomon, 2007). The project also implemented anomaly detection mechanisms to identify any changes from the normal behaviours of the reservoir indicating leaks or other problems, as depicted in Figure 1.8. Even though the project did not involve machine learning or deep learning techniques for anomaly identification, it incorporated classical monitoring techniques for geological security. However, machine learning and deep learning approaches can be implemented on the collected datasets, thereby, identifying the statistical patterns and anomalies. Key aspects of the Sleipner CO2 injection project are:

(1) Monitoring techniques: The injected plume was monitored by geophysical method. Time-lapse gravity and seismic monitoring have proved successful in imaging CO2 plume migration. Datasets from 1999 were quantitatively analysed, and it was found that the seismic signature was consistent with known injected amounts of CO2 (Chadwick, Arts, Eiken, Williamson, & Williams, 2006).

(2) Caprock integrity: There is a 200–300 m impermeable layer of shale cap-rock which prevents injected CO2 from rising to the surface (Kalam, Olayiwola, Al-Rubaii, Amaechi, Jamal, & Awotunde, 2021). Thus, CO2 storage is considered feasible without any leakage from caprock. Stratigraphic, geomechanical, and geochemical factors may affect the caprock integrity.

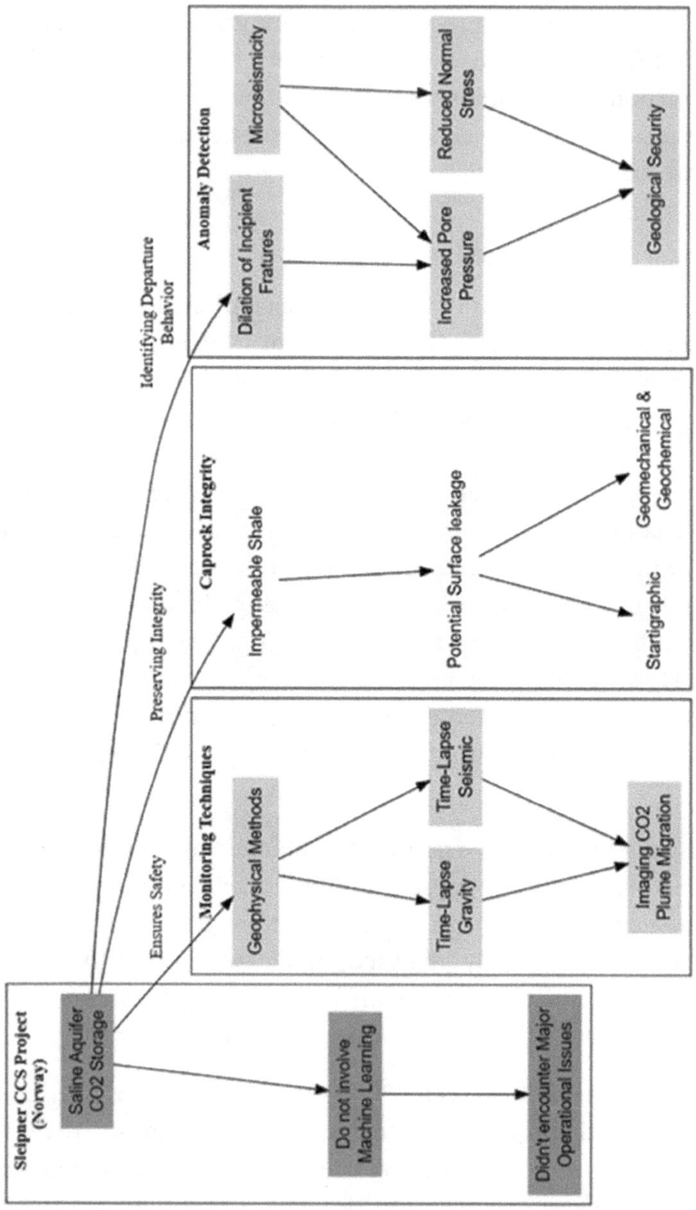

FIGURE 1.8 Sleipner project and importance of anomaly detection.

(3) Anomaly detection: Microseismic studies suggest the injection of CO2 has not trigged any measurable microseismicity; furthermore it builds up the confidence in geological security of CO2 storage (Solomon, 2007). Microseismisity may be induced due to either increased pore pressure or reduced normal stress. Dilation of incipient fractures may also be induced due to high pore pressure.

1.5.2 In Salah CO2 Injection (Algeria)

The In Salah project was started in 2004 and represents an innovative approach to carbon sequestration by adopting a long-term CO2 storage while producing natural gas as a means of enhanced gas recovery (Vasco et al., 2018). The project's monitoring and modelling activities have produced important information about the process of injecting CO2 into a reservoir and its likely impacts on a nearby ecosystem. Within this project, anomaly detection was crucial in identifying any departure from anticipated behaviour in the reservoir that might indicate problems with the CO2 injection (Jones et al., 2011; Morris, Hao, Foxall, & McNab, 2011; Ringrose et al., 2013). Key aspects of the In Salah CO2 injection project are:

(1) Monitoring techniques: This resulted in the establishment of seasonal variations and baseline (background) CO2 values in addition to diverse types of monitoring techniques that recognise any anomalies (Ringrose et al., 2013). Such techniques include applying digital image speckle reduction and analysis (using DInSAR, differential interferometry synthetic aperture radar data) to detect any surface deformation caused by CO2 injection in addition to soil gas and flow data (Jones et al., 2011).

(2) Caprock integrity: The focus of this project was primarily on monitoring and modelling caprock integrity to ensure long-term CO2 storage in deep saline aquifer (Vasco et al., 2018). Notably, the investigation of the injection-induced mechanical deformation of the reservoir and identification of baseline values (normal values) for caprock integrity are among the significant elements of this project (Morris et al., 2011).

(3) Anomaly detection: The aim of In Salah CCS project was to find signs like subsidence that could indicate problems with CO2 injections (Morris et al., 2011). At present, there are only a few conspicuous

signs including slightly increased levels of carbon dioxide compared to normal levels (Ringrose et al., 2013).

1.5.3 Petra Nova Project (United States)

The Petra Nova Project, located at the WA Parish Electric Generating Station in the United States, is a commercial-scale post-combustion carbon capture project which involves all the three steps of the CCS project—capturing, transporting, and injecting CO2—as a means of enhanced oil recovery (EOR) into the West Ranch oilfield to increase oil output (Gul). It was supposed to capture about 5,200 short tons of CO2 per day at maximum capacity with 92% captured. In this project, anomaly detection was very important since it facilitated the identification of any abnormal behaviours of the reservoir (refer to Figure 1.9) that would indicate problems with CO2 injection (Gul). The following are monitoring and anomaly detection efforts made in Petra Nova Project:

(1) Monitoring techniques: The research used a variety of monitoring methods such as fluid sampling and analysis, geophysical logging, and soil gas data for detecting any anomalous behaviour. Likewise, data was also collected from above-zone and in-zone monitoring wells equipped with bottom-hole pressure and temperature sensors to follow the pressure and temperature conditions in the reservoir (Gul).

(2) Caprock integrity: Brittle fossiliferous Anahuac shale act as a secondary seal which will prevent upward migration of injected CO2. It is 120 ft thick in the West Ranch area and has 0.0006–0.0026 md permeability. Some of the potential surface leakage pathways are: diffusion through Anahuac shale, leakage through faults and fractures, and leakage through natural and induced seismicity (Env).

(3) Anomaly detection: For example, if there is subsidence, then that could mean that there is an issue with the injection of CO2. The amounts and isotopic compositions of naturally-occurring gases were also determined by the models to identify the sources of soil gas anomalies (Gul).

Regardless of many hindrances and a final failure of the project to reach carbon capture targets, the monitoring and anomaly detection efforts offer

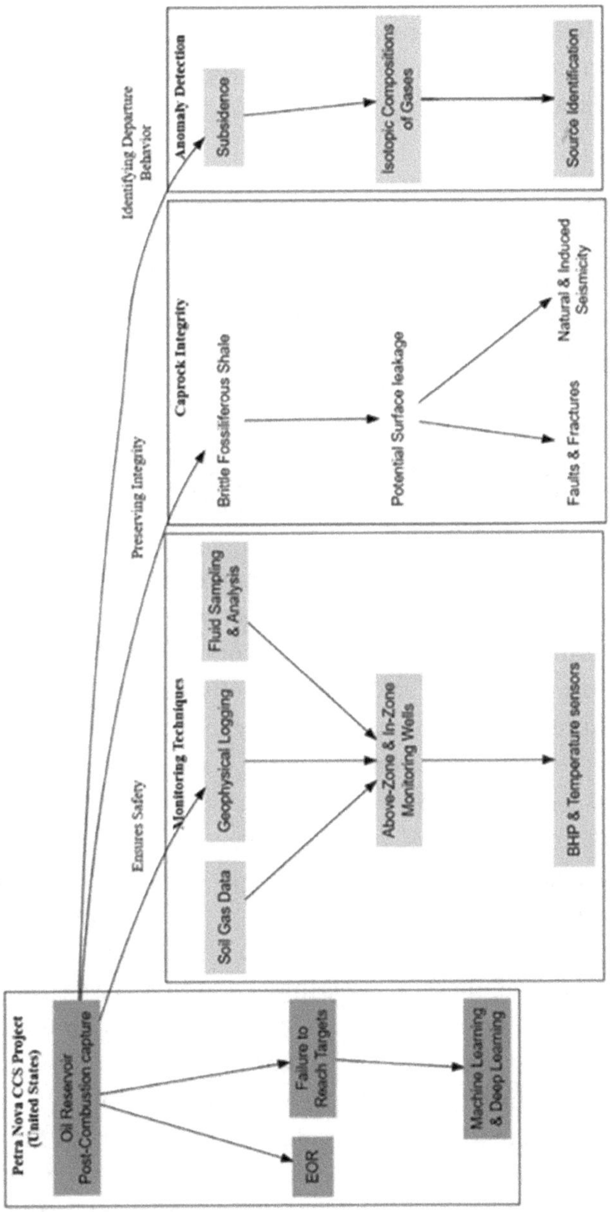

FIGURE 1.9 Anomaly detection in Petra Nova project.

crucial insights into the importance of these processes in assuring the effectiveness of carbon sequestration initiatives (Ins). By incorporating high-end machine learning and deep learning procedures into upcoming projects, there is huge potential to amplify the efficacy of anomaly detection methods and reinforce continuous actions to mitigate climate change by means of carbon sequestration.

1.6 SUMMARY AND CONCLUSION

Anomaly detection is a crucial element in the pursuit of safe, efficient, and effective carbon sequestration projects. To maintain environmental security and mitigate climate change, GCS necessitates close observation and quick action in the event of anomaly detection. Future anomaly detection in carbon sequestration will be secured by ongoing technological and research advancements, international cooperation, and adherence to the best practices. As anomaly detection makes a major contribution to the worldwide effort in preventing climate change, its role in maintaining the integrity and efficiency of carbon storage should not be underestimated. There is active research being done on geologic carbon sequestration to lower greenhouse gas emissions. Authorities, operators, and the public have serious concerns about making sure that CO2 leaks are promptly detected and mitigated. One technique for early detection of CO2 leakage to subterranean drinking water sources is the pressure-anomaly inversion approach.

Because of its early detection capacity, deep subsurface, pressure-based monitoring is an important MMV tool for many GCS sites. In addition to lowering hazards, optimising monitoring network designs also lowers GCS site long-term operating expenses. Because of pressurisation, brine leakage into AZMI may affect bigger areas than CO2 plume leaks and negatively impact drinking water aquifers. Tmax is important from a practical and regulatory standpoint. Only when Tmax is clearly defined and integrated into a site-specific risk management plan can the optimisation problem emerge.

With the use of different neural network architectures for anomaly detection, processes like MFNN, CNN, LSTM, and ConvLSTM, anomaly detectors based on neural networks, can be helpful in providing early warning for CO2 leaks, which can be an invaluable tool to prevent further leakage and improve the safety of carbon sequestration projects. We can create a ranking system based on a variety of quantitative and qualitative factors, such as model scalability, MSE, and practicality in a field deployment scenario, to compare all the models. LSTM is typically placed highest, followed by CNN and MFNN. The MFNN architecture may cause occasional false alarms, but it works effectively with test, validation, and training data. Overall, the LSTM architecture works the best, but the CNN does a great job of capturing features that resemble sinusoidal waveforms. Although the hybrid architecture ConvLSTM offers the best of both worlds, it may have limitations due to its lengthy computation durations. The dynamic nature of carbon sequestration projects is influenced by intricate geology and operational parameters. The efficacy of machine learning methods, such as deep learning, is contingent upon the availability of training data. Deep learning-based anomaly detectors are an affordable and effective substitute for human resources since deep learning architectures are flexible enough to adapt to new datasets and can be employed in a wide range of monitoring projects and field settings. For CO2 leakage detection from a CO2-EOR site, a spatiotemporal ConvLSTM neural network model can be employed. The capacity of the machine learning approach to combine temporal and spatial information makes abnormalities easier to identify. Convolutional neural nets can learn spatial information from the input images. Temporal characteristics can be learned using LSTM models. When a predictive reservoir model is required, accurate leak detection depends on a complete characterisation of the reservoir. According to certain studies, anomalies are easier to find following injection into a reservoir with limited permeability. The ability to anticipate nominal pressure patterns also depends on the availability of operation data, such as injection rate.

ACKNOWLEDGEMENT(S)

The authors would like to thank the reviewers and editors for the feedback.

REFERENCES

1. Amid, S., & Mesri Gundoshmian, T. (2017). Prediction of output energies for broiler production using linear regression, ANN (MLP, RBF), and ANFIS models. *Environmental Progress & Sustainable Energy*, 36(2), 577–585.

2. Benson, S. (2007). Monitoring GS of carbon dioxide. In *Carbon capture and GS: Integrating technology*. Monitoring, and Regulation Blackwell Scientific Publishing.
3. Birkholzer, J. T., & Zhou, Q. (2009). Basin-scale hydrogeologic impacts of CO2 storage: Capacity and regulatory implications. *International Journal of Greenhouse Gas Control*, 3(6), 745–756.
4. Bontemps, L., Cao, V. L., McDermott, J., & Le-Khac, N.-A. (2016). Collective anomaly detection based on long short-term memory recurrent neural networks. In *Future data and security engineering: Third International Conference, FDSE 2016, Can Tho City, Vietnam, November 23–25, 2016, proceedings 3* (pp. 141–152). Springer.
5. Buscheck, T. A., Friedmann, S. J., Sun, Y., Chen, M., Hao, Y., Wolery, T. J., & Aines, R. D. (2012). Aines. Active CO2 reservoir management for CO2 capture, utilization, and storage: An approach to improve CO2 storage capacity and to reduce risk. In *Carbon Management Technology Conference* (pp. CMTC–151746). CMTC.
6. Callegari, C., Giordano, S., & Pagano, M. (2014). Neural network based anomaly detection. In *2014 IEEE 19th International Workshop on Computer Aided Modeling and Design of Communication Links and Networks (CAMAD)* (pp. 310–314). IEEE.
7. Celia, M. A., Nordbotten, J. M., Court, B., Dobossy, M., & Bachu, S. (2011). Field-scale application of a semi-analytical model for estimation of CO2 and brine leakage along old wells. *International Journal of Greenhouse Gas Control*, 5(2), 257–269.
8. Chadwick, A., Arts, R., Eiken, O., Williamson, P., & Williams, G. (2006). Geophysical monitoring of the CO2 plume at Sleipner, North Sea. In *Advances in the geological storage of carbon dioxide: International approaches to reduce anthropogenic greenhouse gas emissions* (pp. 303–314). Springer.
9. Che, Z., Purushotham, S., Cho, K., Sontag, D., & Liu, Y. (2018). Recurrent neural networks for multivariate time series with missing values. *Scientific Reports*, 8(1), 6085.
10. Chen, B., Harp, D. R., Lin, Y., Keating, E. H., & Pawar, R. J. (2018). Geologic CO2 sequestration monitoring design: A machine learning and uncertainty quantification based approach. *Applied Energy*, 225, 332–345.
11. Cihan, A., Zhou, Q., & Birkholzer, J. T. (2011). Analytical solutions for pressure perturbation and fluid leakage through aquitards

and wells in multilayered-aquifer systems. *Water Resources Research*, 47(10), 1–17.

12. Donahue, J., Hendricks, L. A., Guadarrama, S., Rohrbach, M., Venugopalan, S., Saenko, K., & Darrell, T. (2015). Long-term recurrent convolutional networks for visual recognition and description. In *Proceedings of the IEEE Conference on Computer Vision and Pattern Recognition* (pp. 2625–2634). IEEE.
13. Ebigbo, A., Class, H., & Helmig, R. (2007). CO2 leakage through an abandoned well: Problem-oriented benchmarks. *Computational Geosciences*, 11, 103–115.Environmental protection agency, u. (2021). Technical review of subpart RR MRV plan for Petra Nova west ranch unit.
14. Fessenden, J. E., Clegg, S. M., Rahn, T. A., Humphries, S., & Baldridge, W. (2010). Novel MVA tools to track CO2 seepage, tested at the ZERT controlled release site in Bozeman, MT. *Environmental Earth Sciences*, 60, 325–334.
15. Gasda, S. E., Bachu, S., & Celia, M. A. (2004). Spatial characterization of the location of potentially leaky wells penetrating a deep saline aquifer in a mature sedimentary basin. *Environmental Geology*, 46, 707–720.
16. Gu, J., Wang, Z., Kuen, J., Ma, L., Shahroudy, A., Shuai, B., Liu, T., Wang, X., Wang, G., Cai, J., et al. (2018). Recent advances in convolutional neural networks. *Pattern Recognition*, 77, 354–377.Gulf Coast Carbon Center, U. of A. (2020). Petra Nova Carbon Capture and Storage Monitoring — Gulf Coast Carbon Center [WWW Document], howpublished = https://gccc.beg.utexas.edu/research/petra-nova, note. Accessed: 2023-12-21.
17. Haszeldine, R. S. (2009). Carbon capture and storage: How green can black be? *Science*, 325(5948), 1647–1652.
18. Hochreiter, S., Bengio, Y., Frasconi, P., & Schmidhuber, J., et al. (2001). Gradient flow in recurrent nets: The difficulty of learning long-term dependencies. IEEE.
19. Hochreiter, S., & Schmidhuber, J. (1997). Long short-term memory. *Neural Computation*, 9(8), 1735–1780.
20. Hovorka, S. D., Meckel, T. A., & Trevino, R. H. (2013). Monitoring a large-volume injection at Cranfield, Mississippi—Project design and recommendations. *International Journal of Greenhouse Gas Control*, 18, 345–360.

21. Humez, P., Audigane, P., Lions, J., Chiaberge, C., & Bellenfant, G. (2011). Modeling of CO2 leakage up through an abandoned well from deep saline aquifer to shallow fresh groundwaters. *Transport in Porous Media*, 90(1), 153–181.
22. Hussin, F., Rahim, S. A. N. M., Hatta, N. S. M., Aroua, M. K., & Mazari, S. A. (2023). A systematic review of machine learning approaches in carbon capture applications. *Journal of CO2 Utilization*, 71, 102474.Institute of energy economics and financial analysis. (2022). The ill-fated Petra Nova CCS project: NRG energy throws in the towel [www document]. https://ieefa.org/resources/ill-fated-petra-nova-ccs-project-nrg-energy-throws-towel. Accessed: 2023-12-21.
23. Jenkins, C., Chadwick, A., & Hovorka, S. D. (2015). The state of the art in monitoring and verification—Ten years on. *International Journal of Greenhouse Gas Control*, 40, 312–349.
24. Ji, S., Xu, W., Yang, M., & Yu, K. (2012). 3D convolutional neural networks for human action recognition. *IEEE Transactions on Pattern Analysis and Machine Intelligence*, 35(1), 221–231.
25. Jones, D., Lister, T., Smith, D., West, J., Coombs, P., Gadalia, A., Brach, M., Annunziatellis, A., & Lombardi, S. (2011). In Salah gas CO2 storage JIP: Surface gas and biological monitoring. *Energy Procedia*, 4, 3566–3573.
26. Jordan, P. D., Oldenburg, C. M., & Nicot, J.-P. (2011). Estimating the probability of CO2 plumes encountering faults. *Greenhouse Gases: Science and Technology*, 1(2), 160–174.
27. Jung, R. E., Mead, B. S., Carrasco, J., & Flores, R. A. (2013). The structure of creative cognition in the human brain. *Frontiers in human neuroscience*, 7, 330.
28. Kalam, S., Olayiwola, T., Al-Rubaii, M. M., Amaechi, B. I., Jamal, M. S., & Awotunde, A. A. (2021). Carbon dioxide sequestration in underground formations: Review of experimental, modeling, and field studies. *Journal of Petroleum Exploration and Production*, 11, 303–325.
29. Kaur, H., Singh, G., & Minhas, J. (2013). A review of machine learning based anomaly detection techniques. *arXiv preprint arXiv:1307.7286*.

30. Kelemen, P., Benson, S. M., Pilorgé, H., Psarras, P., & Wilcox, J. (2019). An overview of the status and challenges of CO2 storage in minerals and geological formations. *Frontiers in Climate*, 1, 482595.
31. Khan, A., Sohail, A., Zahoora, U., & Qureshi, A. S. (2020). A survey of the recent architectures of deep convolutional neural networks. *Artificial Intelligence Review*, 53, 5455–5516.
32. Kopp, A., Binning, P. J., Johannsen, K., Helmig, R., & Class, H. (2010). A contribution to risk analysis for leakage through abandoned wells in geological CO2 storage. *Advances in Water Resources*, 33(8), 867–879.
33. Le Quéré, C., Moriarty, R., Andrew, R. M., Canadell, J. G., Sitch, S., Korsbakken, J. I., Friedlingstein, P., Peters, G. P., Andres, R. J., Boden, T. A., et al. (2015). Global carbon budget 2015. *Earth System Science Data*, 7(2), 349–396.
34. LeCun, Y., & Bengio, Y. (1995). Convolutional networks for images, speech, and time series. In *The handbook of brain theory and neural networks* 3361(10), 1–14. AT&T Bell Laboratories.
35. Lemieux, J.-M. (2011). The potential impact of underground geological storage of carbon dioxide in deep saline aquifers on shallow groundwater resources. *Hydrogeology Journal*, 19(4), 757.
36. Lewicki, J. L., Hilley, G. E., & Oldenburg, C. M. (2005). An improved strategy to detect CO2 leakage for verification of geologic carbon sequestration. *Geophysical Research Letters*, 32(19).
37. Ma, X., & Hovy, E. (2016). End-to-end sequence labeling via bi-directional LSTM-CNNs-CRF. *arXiv preprint arXiv:1603.01354*.
38. Morris, J. P., Hao, Y., Foxall, W., & McNab, W. (2011). A study of injection-induced mechanical deformation at the in Salah CO2 storage project. *International Journal of Greenhouse Gas Control*, 5(2), 270–280.
39. Mozer, M. C. (2013). A focused backpropagation algorithm for temporal pattern recognition. In *Backpropagation* (pp. 137–169). Psychology Press.
40. Nguyen, M., He, T., An, L., Alexander, D. C., Feng, J., Yeo, B. T., & Alzheimer's Disease Neuroimaging Initiative. (2020). Predicting Alzheimer's disease progression using deep recurrent neural networks. *NeuroImage*, 222, 117203.

41. Nordbotten, J. M., Celia, M. A., & Bachu, S. (2004). Analytical solutions for leakage rates through abandoned wells. *Water Resources Research*, 40(4), 1–10.
42. Nordbotten, J. M., Celia, M. A., & Bachu, S. (2005). Injection and storage of CO2 in deep saline aquifers: Analytical solution for CO2 plume evolution during injection. *Transport in Porous Media*, 58, 339–360.
43. Nordbotten, J. M., Celia, M. A., Bachu, S., & Dahle, H. K. (2005). Semianalytical solution for CO2 leakage through an abandoned well. *Environmental Science & Technology*, 39(2), 602–611.
44. Oladyshkin, S., Class, H., Helmig, R., & Nowak, W. (2011). An integrative approach to robust design and probabilistic risk assessment for CO2 storage in geological formations. *Computational Geosciences*, 15, 565–577.
45. Oldenburg, C. M., Bryant, S. L., & Nicot, J.-P. (2009). Certification framework based on effective trapping for geologic carbon sequestration. *International Journal of Greenhouse Gas Control*, 3(4), 444–457.
46. Oldenburg, C. M., & Unger, A. J. (2003). On leakage and seepage from geologic carbon sequestration sites: Unsaturated zone attenuation. *Vadose Zone Journal*, 2(3), 287–296.
47. Pruess, K. (2004). Numerical simulation of CO2 leakage from a geologic disposal reservoir, including transitions from super-to subcritical conditions, and boiling of liquid CO2. *SPE Journal*, 9(02), 237–248.
48. Pruess, K. (2008). Leakage of CO2 from geologic storage: Role of secondary accumulation at shallow depth. *International Journal of Greenhouse Gas Control*, 2(1), 37–46.
49. Ringrose, P., Mathieson, A., Wright, I., Selama, F., Hansen, O., Bissell, R., Saoula, N., & Midgley, J. (2013). The in Salah CO2 storage project: Lessons learned and knowledge transfer. *Energy Procedia*, 37, 6226–6236.
50. Romanak, K., Bennett, P., Yang, C., & Hovorka, S. (2011). Process-based approach to carbon storage leakage detection using soil gas monitoring: Lessons from natural and industrial analogues. In *AGU fall meeting abstracts* (Vol. 2011, pp. H23B–1246). American Geophysical Union.
51. Rutqvist, J., Vasco, D. W., & Myer, L. (2010). Coupled reservoir-geomechanical analysis of CO2 injection and ground deformations

at in Salah, Algeria. *International Journal of Greenhouse Gas Control*, 4(2), 225–230.

52. Rynkiewicz, J. (2012). General bound of overfitting for MLP regression models. *Neurocomputing*, 90, 106–110.
53. Saif, M., Kiran, R., Rajak, V. K., & Verma, R. K. (2024). Investigation of an Indian site with mafic rock for carbon sequestration. *ACS Omega*.
54. Sathyanarayana, S. (2014). *A gentle introduction to backpropagation: Numeric insight*. Inc Whitepaper.
55. Seto, C., & McRae, G. (2011). Reducing risk in basin scale CO2 sequestration: A framework for integrated monitoring design. *Environmental Science & Technology*, 45(3), 845–859.
56. Shi, X., Chen, Z., Wang, H., Yeung, D.-Y., Wong, W.-K., & Woo, W.-C. (2015). Convolutional LSTM network: A machine learning approach for precipitation nowcasting. *Advances in Neural Information Processing Systems*, 28, 1–9.
57. Shukla, R., Ranjith, P., Haque, A., & Choi, X. (2010). A review of studies on CO2 sequestration and caprock integrity. *Fuel*, 89(10), 2651–2664.
58. Simonyan, K., & Zisserman, A. (2014). Very deep convolutional networks for large-scale image recognition. *arXiv preprint arXiv:1409.1556*.
59. Sinha, S., Pires De Lima, R., Lin, Y., Sun, A. Y., Symon, N., Pawar, R., & Guthrie, G. (2020). Leak detection in carbon sequestration projects using machine learning methods: Cranfield site, Mississippi, USA. In *SPE annual technical conference and exhibition?* (p. D031S022R007). SPE.
60. Solomon, S. (2007). Carbon dioxide storage: Geological security and environmental issues–case study on the Sleipner gas field in Norway. *Bellona Report*, 128, 1–128.
61. Song, Y., Jun, S., Na, Y., Kim, K., Jang, Y., & Wang, J. (2023). Geomechanical challenges during geological CO2 storage: A review. *Chemical Engineering Journal*, 456, 140968.
62. Stauffer, P. H., Viswanathan, H. S., Pawar, R. J., & Guthrie, G. D. (2009). A system model for geologic sequestration of carbon dioxide. ACS Publications.
63. Sun, A. Y., Lu, J., Freifeld, B. M., Hovorka, S. D., & Islam, A. (2016). Using pulse testing for leakage detection in carbon storage reser-

voirs: A field demonstration. *International Journal of Greenhouse Gas Control*, 46, 215–227.

64. Sun, A. Y., Lu, J., & Hovorka, S. (2015). A harmonic pulse testing method for leakage detection in deep subsurface storage formations. *Water Resources Research*, 51(6), 4263–4281.
65. Sun, A. Y., Lu, J., & Islam, A. (2017). A laboratory validation study of the time-lapse oscillatory pumping test for leakage detection in geological repositories. *Journal of Hydrology*, 548, 598–604.
66. Sun, A. Y., & Nicot, J.-P. (2012). Inversion of pressure anomaly data for detecting leakage at geologic carbon sequestration sites. *Advances in Water Resources*, 44, 20–29.
67. Sun, A. Y., Zeidouni, M., Nicot, J.-P., Lu, Z., & Zhang, D. (2013). Assessing leakage detectability at geologic CO2 sequestration sites using the probabilistic collocation method. *Advances in Water Resources*, 56, 49–60.
68. Vasco, D. W., Bissell, R. C., Bohloli, B., Daley, T. M., Ferretti, A., Foxall, W., Goertz-Allmann, B. P., Korneev, V., Morris, J. P., Oye, V., et al. (2018). Monitoring and modeling caprock integrity at the in Salah carbon dioxide storage site, Algeria. In *Geological carbon storage: Subsurface seals and caprock integrity* (pp. 243–269). Wiley.
69. Viswanathan, H. S., Pawar, R. J., Stauffer, P. H., Kaszuba, J. P., Carey, J. W., Olsen, S. C., Keating, G. N., Kavetski, D., & Guthrie, G. D. (2008). Development of a hybrid process and system model for the assessment of wellbore leakage at a geologic CO2 sequestration site. *Environmental Science & Technology*, 42(19), 7280–7286.
70. Wang, L., Bai, B., Li, X., Liu, M., Wu, H., & Hu, S. (2016). An analytical model for assessing stability of pre-existing faults in caprock caused by fluid injection and extraction in a reservoir. *Rock Mechanics and Rock Engineering*, 49, 2845–2863.www.dnv.com, 2023. dnv.com—when trust matters—DNV [www document]. https://www.dnv.com/. Accessed: 2023-11-10.
71. Yang, Y.-M., Dilmore, R. M., Bromhal, G. S., & Small, M. J. (2018). Toward an adaptive monitoring design for leakage risk–closing the loop of monitoring and modeling. *International Journal of Greenhouse Gas Control*, 76, 125–141.
72. Yao, P., Yu, Z., Zhang, Y., & Xu, T. (2023). Application of machine learning in carbon capture and storage: An in-depth insight from the perspective of geoscience. *Fuel*, 333, 126296.

73. Yilmaz, I., & Kaynar, O. (2011). Multiple regression, ANN (RBF, MLP) and ANFIS models for prediction of swell potential of clayey soils. *Expert Systems with Applications*, 38(5), 5958–5966.
74. Zeidouni, M. (2012). Analytical model of leakage through fault to overlying formations. *Water Resources Research*, 48(12), 1–17.
75. Zeiler, M. D., & Fergus, R. (2014). Visualizing and understanding convolutional networks. In *Computer vision–ECCV 2014: 13th European Conference,* Zurich, Switzerland,*September 6–12, 2014, proceedings, part I 13* (pp. 818–833). Springer.
76. Zhong, Z., & Carr, T. R. (2019). Geostatistical 3D geological model construction to estimate the capacity of commercial scale injection and storage of CO2 in Jacksonburg-Stringtown oil field, West Virginia, USA. *International Journal of Greenhouse Gas Control*, 80, 61–75.
77. Zhong, Z., Sun, A. Y., Yang, Q., & Ouyang, Q. (2019). A deep learning approach to anomaly detection in geological carbon sequestration sites using pressure measurements. *Journal of Hydrology*, 573, 885–894.
78. Zhou, Q., Birkholzer, J. T., Tsang, C.-F., & Rutqvist, J. (2008). A method for quick assessment of CO2 storage capacity in closed and semi-closed saline formations. *International Journal of Greenhouse Gas Control*, 2(4), 626–639.
79. Zhu, G., Zhang, L., Shen, P., & Song, J. (2017). Multimodal gesture recognition using 3-D convolution and convolutional LSTM. *IEEE Access*, 5, 4517–4524.

CHAPTER 2

Intelligent Anomaly Detection in Social Media

Yanwei Xu, Zhiyong Feng, and Schahram Dustdar

2.1 INTRODUCTION

With the development of Web 4.0, social media has become an essential tool for people to establish connections and collaborate. Individuals interact on social media by sharing images, comments, videos, and various forms of content, generating a vast array of diverse user-generated content (Evans, Bratton, & McKee, 2021). While people enjoy the accessibility and convenience of communication on these platforms, malicious actors exploit user-generated content to mine behavioural or interaction patterns for nefarious activities such as cyberbullying, malicious comments, and the dissemination of fraudulent information. These activities have the potential to inflict economic losses or privacy breaches upon users (Yu, Qiu, Wen, Lin, & Liu, 2016). Hence, accurately and promptly detecting such malicious activities is an imperative task to proactively mitigate associated risks and minimise losses.

Generally speaking, anomaly detection is the process of identifying patterns, behaviours, events, or observations that deviate from predefined norms (Ma et al., 2021). In the realm of social media, anomaly detection typically involves identifying users or groups whose behavioural patterns deviate from established norms (Chandola, Banerjee, & Kumar, 2009). For instance, consider Tom and Kim, two users who routinely share recommendations for adventure movies on social media. However, on a particular day, Tom sends Kim a message containing a malicious link, representing a deviation from their typical interactions (as shown in Figure 2.1). Detecting such irregular behaviours becomes crucial at this juncture to mitigate potential

DOI: 10.1201/9781003463559-2

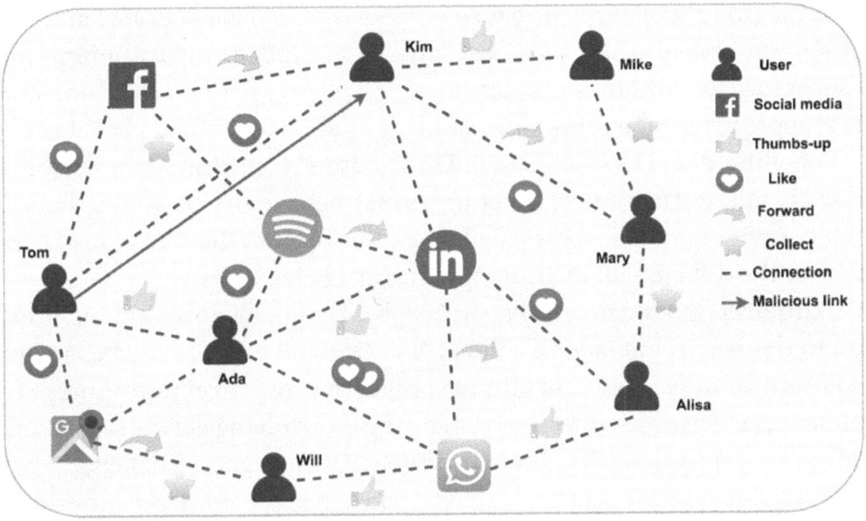

FIGURE 2.1 People connect with each other on social media. Many users express their opinions on social media, for example, with likes, thumbs-up, and forwards. Users also send messages to each other, which may include malicious links.

financial losses for Kim. Specifically, social media platforms like Facebook and Instagram are online communication platforms built on the interactive behaviours between pairs of users or multiple groups. Based on such user interaction-centric features, the majority of connections between users adhere to the social balance theory (Qi, Dou, & Zhang, 2016), encompassing four linking principles: 'A friend of a friend is a friend,' 'A friend of an enemy is an enemy,' 'An enemy of a friend is an enemy,' and 'An enemy of an enemy is a friend.' These four linking criteria offer a potential for detecting anomalous behaviour by observing the linking or non-linking of all users. Simultaneously, it is essential to identify peculiar behaviours through the analysis of user behaviour patterns and interactions for detecting anomalies.

As introduced earlier, social media revolves around user interactions; these interactions are commonly modelled as online social networks, where nodes represent users, and edges represent social relationships between users. Based on the topological structure of online social networks, existing anomaly detection methods mainly focus on two foundational approaches (Yu, Qiu, Wen, Lin, & Liu, 2016): those based on user behaviour and those based on group structures. The former relies on the behavioural characteristics of individual users, while the latter is grounded in partitioning the group structures within social networks. Despite numerous works utilising these approaches for anomaly detection, they still face several challenges:

Labelling data (Li et al., 2021): Due to users' concerns about personal privacy, online users rarely label abnormal behaviours of others; hence, such behaviours remain unlabelled when they occur. These circumstances increase the difficulty of identifying abnormal behaviours.

Temporal dependency (Li et al. 2023b): Daily exchanges among users lead to changes in interaction content (i.e., text and images) and behaviour patterns over time. Consequently, new edges or group structures emerge in online social networks. Effectively capturing the evolving network structure and subsequently detecting anomalous activities presents a challenge.

In this chapter, we present a comprehensive and in-depth investigation into the exploration and analysis of anomaly detection in online social networks. Our main contributions can be summarised as follows:

- An overview of the research background of anomaly detection in online social networks and addressed associated issues.
- A comprehensive hierarchical classification considering static/dynamic user behaviour and static/dynamic network structure evolution.
- An analysis of the practical applications of anomaly detection techniques in real-life scenarios, emphasising commonly used datasets.
- An outline of the significant challenges confronting current anomaly detection methods.
- A set of new research questions for the future.

The remaining structure of this chapter unfolds as follows: Section 2.2 introduces online social networks and anomaly detection separately, paving the way for their integration. Sections 2.3 and 2.4 meticulously discuss the strengths and weaknesses of existing literature on anomaly detection in online social networks. Section 2.5 explores the practical applications of anomaly detection in real-life scenarios. Section 2.6 provides insights into popular datasets in anomaly detection. In Section 2.7, we discuss challenges and key issues in anomaly detection in various real-world scenarios. Finally, Section 2.8 concludes the chapter by summarising the key points and providing an outlook for future research.

2.2 BACKGROUND

In this section, we introduce the research background from two perspectives: the analysis of online social networks and anomaly detection. By separately highlighting the characteristics of these two topics, we aim to provide a comprehensive reference for their integration and further research.

2.2.1 Online Social Network Analysis

In the context of online social networks, a network is defined as a graph comprising nodes and edges. Nodes represent social entities, such as users or groups, while edges represent social relationships between these entities, such as friendships, work connections, and collaborations (Vošner, Bobek, Kokol, & Krečič, 2016). Based on the composition of online social networks, several key characteristics can be outlined:

- **Node attributes**

ID (identifier): Each node possesses a unique identifier, facilitating the distinction between different social entities.

Features/attributes: Nodes may exhibit various features or attributes, encompassing the occupation, age, address, gender, and other relevant characteristics of social entities.

Degree: A node's degree reflects the number of edges connected to it, offering insights into the node's level of connectivity with other nodes. The degree distribution provides a means to delineate the significance of different nodes.

Centrality: This metric emphasises the importance of nodes in the network. The popular metrics contain centrality, closeness centrality, and betweenness centrality.

- **Edge attributes**

Weight: Signifying the strength or intensity of a relationship. Edge weight denotes the degree of association between nodes. Thus, social networks can be categorised into weighted or unweighted based on edge weight.

Direction: Indicates the directionality of the relationship and highlights the asymmetry between two social entities. In general, social networks are modeled as undirected networks (Jiang, Wang, Bhuiyan, & Wu, 2016).

Type: Edges can be classified according to the nature of the social relationship. Based on the type of edge, online social networks can be characterised as co-friendship networks (Li, Zhao, Wang, Zhu, Wu, & Shi, 2017), co-collaborate networks (Blaschke & Hase, 2019), co-authorship networks (Molontay & Nagy, 2019), and email-interaction networks (Nawaz, Khan, & Lee, 2016).

- **Social network theories**

Six degrees of separation theory (Guare, 2016): The six degrees of separation theory refers to the observation that individuals separated by considerable geographic distances may have relatively short social relationship intervals. Stanley Milgram conducted the famous 'Six Degrees of Separation' experiment, suggesting that any two people in the world are connected by an average of six acquaintances. According to a 2011 report by Facebook's data analysis group, the average path length between any two users among approximately 720 million users is only 4.74, and on Twitter, it is 4.67.[1] In other words, within five steps, any two individuals on the network can be connected.

Homophily theory: In social network research, the homophily theory indicates a trend in which nodes in a social network tend to share similarities or common attributes (McPherson, Smith-Lovin, & Cook, 2001). Specifically, the theory suggests that nodes in a social network are more likely to connect with other nodes that share certain attributes. For example, social entities may prefer to form friendships or collaborations with others who share similar interests, activities, or themes. This theory is crucial for understanding information dissemination, opinion formation, and network evolution in social networks.

Scale-free networks (Barabási, 2009): Large-scale social networks exhibit the characteristic that the majority of nodes have few connections, while a minority of nodes have a large number of connections. The network lacks a unified measuring scale and demonstrates heterogeneity, a property referred to as scale-free. Scale-free networks exhibit a power-law distribution in their degree distribution, and social networks adhere to the scale-free characteristics.

Based on the characteristics of social networks, social entities and social relationships are typically modeled as a graph $G=(V, E)$, where $V=v_1$,

1 https://www.brafton.co.uk/news/marketers-separated-from-any-facebook-user-by-just-4-degree/

$v_2,\ldots,v_i, v_j,\ldots,v_n$ represents the set of nodes, and $E = e_{1,2}, e_{1,3},\ldots,e_{i,j}, e_{n-1,n}$ represents the set of edges, where $e_{i,j} = (v_i, v_j)|v_i, v_j \in V$.

Utilising the understanding derived from the composition of social networks, associated theories, and mathematical modelling, social network analysis finds extensive applications in various domains including marketing (Kotler, Burton, Deans, Brown, & Armstrong, 2015), information dissemination, social sentiment detection (Bharne & Bhaladhare, 2023), social recommendation systems (Qi, Wang, Hu, Li, He, & Xu, 2019), and anomaly detection (Chandola, Banerjee, & Kumar, 2009).

2.2.2 Anomaly Detection

Anomaly detection is a process that typically involves identifying data points or samples in a dataset or network that deviate from the established norm. Anomalies are also referred to as outliers, novelties, noise, deviations, or exceptions. In the context of data analysis, anomaly detection aims to flag samples that do not conform to typical patterns or statistical models within the majority of the data. Anomalies can be categorised into the following three types (Chandola, Banerjee, & Kumar, 2009):

Point anomaly: This refers to samples where only a few individual data points are anomalous, while the majority fall within the normal range. For example, in a dataset related to health indicators, most individuals exhibit normal values, but a few may show signs of illness.

Group anomaly: Samples within a group or set exhibit anomalous behaviour, even though the individual instances themselves may not be anomalous. In applications like intrusion or fraud detection, outliers correspond to sequences of multiple data points rather than isolated occurrences. For example, a subset of fraudulent accounts forms a collective anomaly in a social network, where individual nodes within the subset may appear normal, resembling genuine accounts.

Contextual anomaly: Samples are considered anomalous in specific contexts but appear normal in other situations. For instance, a sudden temperature spike or drop at a particular time or a rapid succession of credit card transactions in a specific scenario could be contextual anomalies. Figure 2.2 illustrates the distribution of the three anomalies.

One common characteristic of anomalies is that the detected abnormal samples rarely occur, resulting in a relatively small amount of data.

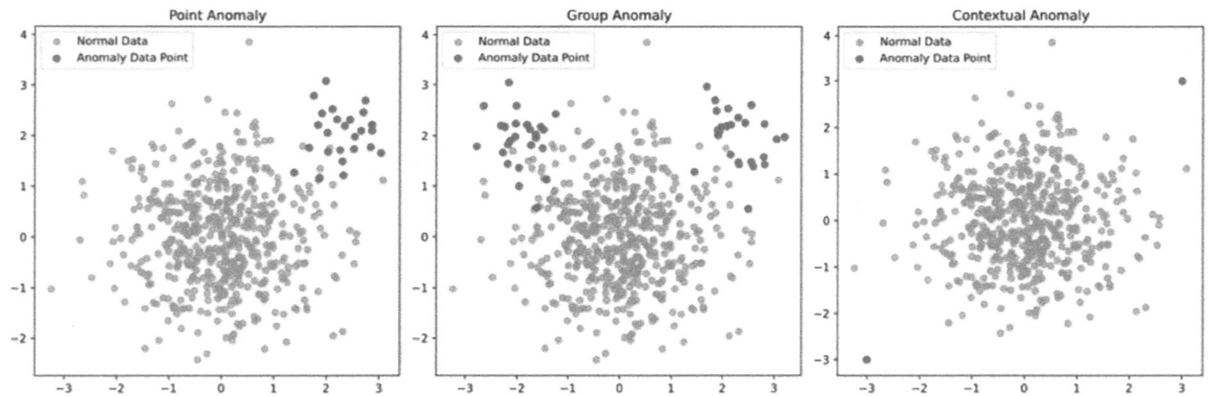

FIGURE 2.2 Different anomaly data points are distributed across three anomaly detection scenarios. In point anomaly, a single data point deviates from the normal sample. In group anomalies, a cluster of abnormal data points diverges from a normal sequence of values. In contextual anomaly, a data point deviates from the range of normal values within a given context.

Therefore, the following models are commonly used to detect irregular values or samples.

- **Statistical models**

In statistical modelling, commonly used measurement methods include *Z*-score, 3α, and others. Specifically, Z-score refers to a method for detecting parameter anomalies in one-dimensional or low-dimensional feature spaces. This technique assumes that the original data follows a Gaussian distribution, and anomalous values are data points in the tails of the distribution, making them far from the mean of the data (Patro & Sahu, 2015). The distance is determined by the normalised data points z_i, calculated using a formula, and depends on the threshold Z_t set for anomalies. Assume that the dataset is $X = x_1, x_2, \ldots, x_m$,

$$Z_i = \frac{x_i - \mu}{\sigma} \tag{2.1}$$

where $x_i \in X$ denotes the data point in the dataset, μ denotes the average score of the data points, and σ denotes the standard deviation of all data points. Then after normalisation, the outliers are also normalised and their absolute values are greater than Z_t.

$$|Z_i| > Z_t \tag{2.2}$$

The Z_t value is typically set to 2.5, 3.0, and 3.5, which conforms to most cases of anomaly detection (Patro & Sahu, 2015).

- **Classification model**

The one-class classification algorithm is commonly used in classification models, where a small portion of the training samples is considered anomalous. It trains an extremely strict decision boundary, encompassing the majority of samples identified as normal, and uses this boundary as the measurement standard.

The support vector machine (SVM) is a supervised learning model used for classification and regression analysis (Wang & Hu, 2005). In classification problems, SVM aims to find an optimal hyperplane to separate samples of different classes while maximising the margin on both sides of the hyperplane. One-class SVM is a variant of the SVM designed for anomaly detection problems where there is only one class (Ruff et al., 2018). Its main objective is to identify anomalous samples that are significantly different from the normal class. The core idea is that the one-class SVM attempts to enclose normal samples in a high-dimensional space, forming a 'wrapper.' During testing, it evaluates whether a new sample is within this wrapper, and if not, it is considered anomalous. The one-class SVM optimisation problem is formulated as follows.

$$\begin{aligned} \underset{w,\xi,\rho}{\text{minimisze}}\ & \frac{1}{2}\|w\|^2 - \rho + \frac{1}{\nu m}\sum_{i=1}^{m}\xi_i \\ \text{subject to}\ & \langle w, \varphi(x_i)\rangle \geq \rho - \xi_i, i = 1, ..., m \\ & \xi_i \geq 0, i = 1, ..., m \end{aligned} \tag{2.3}$$

where w is the weight vector of the hyperplane, ξ_i is the slack variable, ρ is the distance to the hyperplane, ν is a user-defined parameter controlling the proportion of outliers, m is the number of training samples, and $\varphi(x)$ is the feature mapping of the input sample x. Figure 2.3 illustrates visually the results of the one-class SVM algorithm to detect anomalous data points.

- **Linear models**

Commonly used linear model methods, such as principal component analysis (PCA), are applied to reduce the dimensionality of the original features (Kherif & Latypova, 2020). The goal is to construct a new feature space that maps the original data to a lower-dimensional space while maximising the preservation of the original data's characteristics (measured by data covariance). The principle is to speed up the data processing and alleviate the 'curse of dimensionality.' For the original data $X = x_1, x_2, \ldots, x_m$, the following transformations are applied.

$$C = \frac{1}{m-1}\hat{X}^T\hat{X} \tag{2.4}$$

$$Z = \hat{X}U \tag{2.5}$$

where the mean vector of the data is computed by $\mu = \frac{1}{m}\sum_{i=1}^{m} x_i$ and the data matrix is normalised by $\hat{X} = X - \mu$. C denotes the covariance matrix. The eigenvalues $[\lambda_1, \lambda_2, ..., \lambda_n]$ are obtained by eigenvalue decomposition of the covariance matrix and the corresponding eigenvector $[v_1, v_2, ..., v_n]$. Then, the eigenvalues are sorted according to their rankings in terms of size, and the eigenvectors corresponding to the top k eigenvalues are selected to

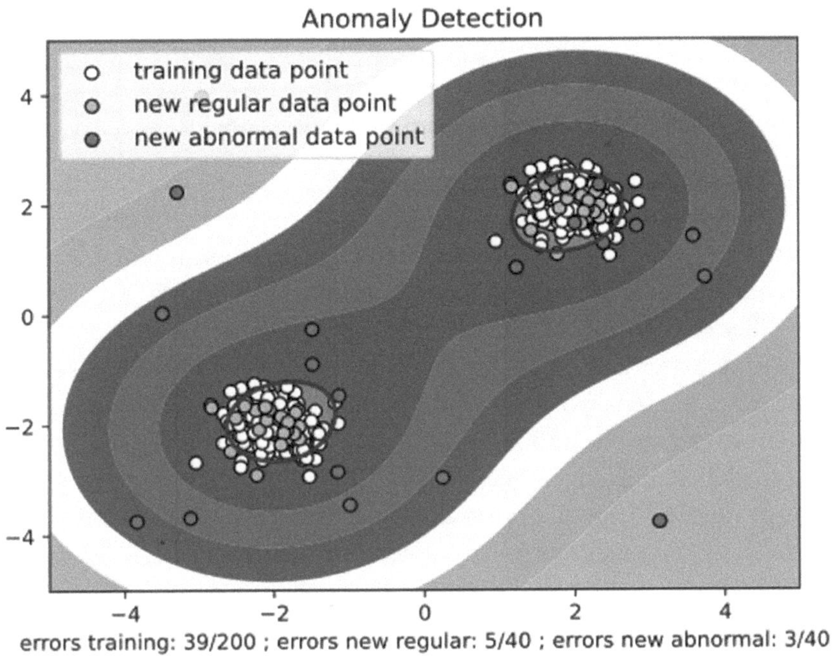

FIGURE 2.3 An example of a one-class SVM detecting anomalous data. By using a one-class SVM, the data can be hierarchically classified, with training data point, new regular data point, and new anomalous data point.

form the projection matrix *U*. By multiplying the data matrix with the projection matrix, the dimensionality reduced data matrix *Z* is obtained. The data is then projected into a *k*-dimensional space, obtaining a reduction in dimensionality. PCA is capable of effectively reducing the computational overhead of the algorithm and is completely parameter-free.

Leveraging background knowledge in the field of online social network analysis enhances our understanding of how anomaly detection is applied within these networks. Consequently, this deepened understanding contributes to a clearer comprehension of the overall impact of anomaly detection in online social networks.

2.3 BEHAVIOUR-BASED ANOMALY DETECTION

Behaviour-based anomaly detection methods in social networks primarily focus on the individual behavioural characteristics or patterns of a single user. Subsequently, they aim to identify anomalous activities between pairs of users.

2.3.1 Static Behaviour-Based Anomaly Detection

People engage in mutual communication, sharing, and collaboration on social media. They post reviews, give likes, and bookmark content to express their attitudes and emotions. These interactions contain implicit traces of user behaviour and patterns (Donta, Srirama, Amgoth, & Annavarapu, 2022; Miller et al., 2016). Mining this behavioural data not only contributes to providing better and more personalised services for the public but also aids in quickly identifying potential malicious activities and mitigating potential dangers to the public.

Given that online social networks can be modeled as graphs with statistical properties, Chaudhary et al. (Chaudhary, Mittal, & Arora, 2019) posit that malicious users prefer to connect with normal users. Representing users as nodes with associated node attributes, they propose three hypotheses: a malicious user possesses a higher degree, centrality, and closeness centrality. Nodes are represented in the adjacency matrix of the graph, and graph neural networks are utilised to obtain node features. Based on thresholds derived from the three criteria, anomalous nodes are detected. Considering the nuanced differences between anomalous and normal nodes, Khamparia et al. (Khamparia, Pande, Gupta, Khanna, & Sangaiah, 2020) introduce a multi-level framework for detecting anomalous nodes in social networks. Commencing with the feature extraction process, they utilise hybrid anomaly detection techniques to scrutinise node characteristics, including out-degree/in-degree distribution, clustering coefficient distribution, and eigenvalue distribution. This comprehensive analysis allows for the assessment of the distribution of anomalous nodes.

Taking an alternative approach, a neural network framework is modeled based on user interaction patterns, utilising metapaths to consider the distribution of user connections in the network (Xu et al., 2023). From a graph theory perspective, the statistical analysis of user features overlooks the context information centred around individual users. User behaviour is influenced by the specific contexts in which interactions take place to a certain extent. *TargetVue* (Cao, Shi, Lin, Lu, Lin, & Lin, 2015) is introduced to visualise and analyse contextual information, providing an intuitive representation of user behaviour patterns during the interaction process. This framework integrates various contextual perspectives, such as behavioural features, text features, interaction features, and network features, to identify high-level user characteristics. Additionally, Z-scores are employed to compare the distinctiveness between centralised users and normal users.

The aforementioned methods, primarily based on static behaviour, largely leverage statistical approaches, predominantly constructed upon graph properties (e.g., nodes and edges). However, these approaches tend to overlook the dynamic nature of user-centric interaction patterns that evolve across multiple time periods.

2.3.2 Dynamic Behaviour-Based Anomaly Detection

In the realm of dynamic online social networks, where user interactions constantly shift, behavioural patterns undergo continuous changes over time (Dustdar, Pujol, & Donta, 2022; Donta, Sedlak, Casamayor Pujol, & Dustdar, 2023). This dynamic nature, characterised by the frequent addition and removal of users and relationships, poses an increased level of complexity when mining temporal patterns.

Li et al. (Li, Sun, Xu, & Li, 2017b) conceptualised a series of sequential temporal snapshots $T_i = \{t_1, t_2, ..\}(t_i \in T_i)$ as an evolving social network, with the objective of uncovering temporal trends within each snapshot. To enhance precision, each snapshot network was meticulously labelled as 0 or 1. After defining these temporal patterns, the one-class SVM algorithm was employed to discern anomalous temporal patterns.

Building upon the intuitive premise that the behaviour of a normal user should conform to the predominant behaviour observed across users, Viswanath et al. (Viswanath et al., 2014) segmented users' continuous behaviour over a long period into multiple time windows. The integration of temporal and spatial features facilitated a comprehensive understanding of the evolution of user behaviour. Leveraging PCA techniques, the conversion of multi-scale features into low-dimensional counterparts effectively mitigated the time and computational complexities associated with mining multiple temporal segments. Given that malicious content is often orchestrated by computers, the *Act-M* model (Bandara, Pezeshki, & Anura, 2010), which is grounded in the posting arrival times on influential social media platforms like Reddit and Twitter, defines four distinctive behavioural patterns, including heavy-tails, to identify patterns indicative of automated or bot-like activity. In contrast to custom temporal patterns, records from distinct time periods are traceable and exhibit logical coherence. Kong et al. (Kong, Jones, & Belta, 2016) devised signal temporal logic (STL) inference formulas to encapsulate normal behaviour within a defined range. The logical structure introduced by STL enhances interpretability of behaviour, avoiding abstraction into high-dimensional feature spaces. Utilising the robust feature representation capabilities offered by LSTM networks, Xu et al. (Xu et al., 2023) proposed a comprehensive framework that is capable of feature extraction for user behaviours in different time periods. Further, the combination with the multiple attention mechanism further enhances the model's ability to discriminate the importance of different behavioural patterns in different time intervals.

In the evolving landscape of online social networks, user behaviour interactions are constantly changing. Detecting individual user behaviour features based on temporal and spatial characteristics is localised, neglecting the measurement of whether users conform to the overall evolving patterns of the entire network from a global perspective.

2.4 STRUCTURE-BASED ANOMALY DETECTION

In online social networks, diverse structures emerge from the intrinsic connections and divisions among users, giving rise to various patterns such as ego networks, triadic closure (Yu, Qiu, Wen, Lin, & Liu, 2016), and small-world structures (Guare, 2016). These distinct structures reflect the diverse patterns of behavioural interactions among users. From a structural perspective, they contribute to the detection of anomalous behaviours that deviate from the typical patterns exhibited by the majority of users.

2.4.1 Static Structure-Based Anomaly Detection

On Twitter, users exhibit diverse social relationships and carry personal attribute information such as age, gender, and location. Typically, such online social networks, encompassing user attributes, are characterised as attribute graphs. Due to incorrect links and inconsistent attribute information between anomalous nodes and normal nodes, the attribute graph exhibits a diversity of anomalies. *tGCN* (Luo et al., 2022) has been proposed for community-based anomaly detection, considering the perspectives of local, global, and structural aspects. An autoencoder is employed to encode the modular matrix of the graph, further aggregating neighbour nodes through GCN layers to capture local and global anomalous nodes. A gateway mechanism is introduced into the GCN layer to fuse structural anomaly features. Finally, *tGCN* outputs features indicative of anomalies.

Considering that incorporating prior knowledge into anomaly detection may fail to detect genuine anomalies, *GraphUCB* (Ding, Li, & Liu, 2019) adopts an interactive approach between humans and models. This method employs a multi-arm bandit framework to provide feedback on anomalous node features to human experts, optimising the *GraphUCB* strategy through multiple rounds. Apart from graph attribute representation, departing from the distribution of graph data, Kopp et al. (Kopp, Grill, & Kohout, 2018) create community-based subnetworks based on similar node behaviours. The interaction between two subnetworks is judged by examining whether they share common subnetworks, thus determining their similarity. Jaccard similarity and cosine similarity are utilised to assist in detecting anomalies in individual nodes and communities.

The static network structure serves as a representation of the shared subnetwork within groups, distributed betweenness, yet it fails to account for the dynamic structural evolution between individual nodes and groups, as well as among different groups. The capture of evolving network structures across multiple time periods proves instrumental in effectively delineating the comprehensive behavioural characteristics of nodes and groups.

2.4.2 Dynamic Structure-Based Anomaly Detection

Social networks, especially on social media platforms, are inherently dynamic. The addition of new users and the departure of old users contribute to the evolution of the network's topological structure.

Typically, an evolving social network is represented as a time-varying graph $G = G_1, G_2, \ldots, G_T$. Leveraging the graph's characteristics and adopting a microscopic perspective (i.e., focusing on nodes), it is assumed that the evolution patterns of nodes (i.e., addition and removal of edges) align with the majority of nodes' evolution patterns. Wang et al. (Wang, Wu, Hu, & Wu, 2019) transform node anomalous evolution patterns for application in link prediction detection methods. They introduce a matrix perturbation assessment (MPA) method to reflect structural changes in the graph, confirming the distinctiveness of node behaviour vectors. The connectivity between nodes within a community structure is different from the connectivity between nodes belonging to different communities. Simultaneously, nodes within two communities may exhibit partial overlap, e.g., a Twitter user belonging to both student and family communities. A model associated with community affiliations is proposed (Baingana & Giannakis, 2015), capable of identifying time-aware community structures and detecting anomalous node behaviour. Mining shared community affiliations among nodes is achieved through linear transformations of an adjacency matrix.

Different from the traditional linear transformation of matrices, network embedding proves effective in learning low-dimensional embeddings of nodes while capturing and preserving network structure. *NetWalk* (Yu, Cheng, Aggarwal, Zhang, Chen, & Wang, 2018) takes the initial step of encoding nodes to obtain clique embeddings from the dynamic network. Utilising clustering-based deep autoencoders, it acquires globalised representations for the sustained detection of dynamic network anomalies. Acknowledging the frequent occurrence of false positives in existing anomaly detection models, where normal behavioural deviations are mistakenly identified as anomalies, ACOBE (Yuan, Choo, Yu, Khalil, & Zhu, 2021) integrates individual user and group-related behavioural signals across multiple time windows. Fully-connected autoencoders are employed to uncover compound behavioural features, generating a sequentially ordered list of anomalous users. To further mitigate the rate of erroneous anomaly alerts, a two-stage anomaly detection framework (TSAD) is proposed (Jiang & Liu, 2022). It relies on a bipartite graph composed of a series of time slices and community similarity scores. By incorporating the evolving characteristics of communities, TSAD extracts anomalous evolution paths tonal m identify exceptional nodes through normal evolutionary pathways. As depicted in Table 2.1, we conducted a comparative analysis between traditioethods and machine learning-based approaches.

2.5 DATASET AND CONFIGURATIONS

The availability of effective datasets is indispensable for validating the efficacy of anomaly detection techniques and theories. As shown in Table 2.2, we gathered and analysed datasets commonly used for the work of anomaly detection in online social networks. We considered key factors such as type, the number of nodes, the number of edges, and uniform resource

TABLE 2.1 Comparison of traditional and machine learning-based methods for anomaly detection in online social networks

Category	Reference	Traditional approach				Machine learning-based approach		
		Degree	Cluster	One-class SVM	PCA	GNN	GCN	Autoencoder
Behaviour-based approach	Chaudhary, Mittal, and Arora (2019)	✓	✓	✗	✗	✓	✗	✗
	Khamparia, Pande, Gupta, Khanna, and Sangaiah (2020)	✓	✓	✗	✗	✗	✗	✗
	Cao, Shi, Lin, Lu, Lin, and Lin (2015)	✓	✓	✗	✗	✗	✗	✗
	Li, Sun, Xu, & Li, 2017	✓	✓	✓	✓	✗	✗	✗
	Viswanath et al. (2014)	✓	✓	✗	✓	✗	✗	✗
	Kong, Jones, and Belta (2016)	✓	✓	✗	✗	✗	✗	✗
Structure-based approach	Luo et al. (2022)	✓	✓	✗	✗	✓	✓	✗
	Yu, Cheng, Aggarwal, Zhang, Chen, and Wang (2018)	✓	✓	✗	✗	✗	✗	✓
	Jiang and Liu (2022)	✓	✓	✗	✗	✗	✗	✗
	Yuan, Choo, Yu, Khalil, and Zhu (2021)	✓	✓	✗	✗	✓	✗	✓
	Ding, Li, and Liu (2019)	✓	✓	✗	✗	✗	✗	✓
	Kopp, Grill, and Kohout (2018)	✓	✓	✗	✗	✗	✗	✗
	Ding, Li, and Liu (2019)	✓	✓	✗	✗	✗	✗	✓
	Kopp, Grill, and Kohout (2018)	✓	✓	✗	✗	✗	✗	✗
	Wang, Wu, Hu, and Wu (2019)	✓	✓	✗	✗	✗	✗	✗
	Baingana and Giannakis (2015)	✓	✓	✗	✗	✗	✗	✗

TABLE 2.2 Popular datasets for anomaly detection in online social networks

Dateset	Type	Nodes	Edges	URL
BlogCatalog	Blog, groups, labelled	10,312	333,983	https://figshare.com/articles/dataset/BlogCatalog_dataset/11923611
Flickr	Image, labelled	105,938	2,316,948	https://snap.stanford.edu/data/web-flickr.html
Twitter	Post, labelled	81,306	1,768,149	https://snap.stanford.edu/data/ego-Twitter.html
Google+	Profiles, labelled	107,614	13,673,453	https://snap.stanford.edu/data/ego-Gplus.html
Facebook	Posts, labelled	4,039	88,234	https://snap.stanford.edu/data/ego-Gplus.html
TwitterSecurity	Temporal, post	-	-	http://odds.cs.stonybrook.edu/twittersecurity-dataset/
EnronInc	Temporal	80,884	-	https://odds.cs.stonybrook.edu/enroninc-dataset/
NYTNews	News	320,000	-	https://odds.cs.stonybrook.edu/nytnews-dataset/
Hacker-News	News, comments	-	-	https://www.kaggle.com/datasets/hacker-news/hacker-news-posts
Reddit	Subreddits	55,863	858,490	https://snap.stanford.edu/data/soc-RedditHyperlinks.html
Enron	Email	150	-	https://www.cs.cmu.edu/ ./enron/
DARPA	Network traffic, audit logs	-	-	https://www.ll.mit.edu/r-d/datasets/1998-darpa-intrusion-detection-evaluation-dataset
Telecom	Temporal, tele-comservices	-	-	https://www.kaggle.com/datasets/navins7/telecommunications
UCI Messages	Messages	1,899	13,838	http://konect.cc/networks/opsahl-ucsocial/
Stack-overflow	Questions, anwers	8,755	-	https://archive.org/details/stackexchange
Orkut	Friendship, labelled	3,072,441	117,185,083	https://snap.stanford.edu/data/com-Orkut.html
YelpChi	Reviews	38,063	-	https://odds.cs.stonybrook.edu/yelpchi-dataset/
T-Finance	Transaction	39,357	21,222,543	https://drive.google.com/drive/folders/1DKO8edxNJ3UUWq-w2f6mLlKecr1itEUr
T-Social	Social	5,781,065	73,105,508	https://drive.google.com/drive/folders/1DKO8edxNJ3UUWq-w2f6mLlKecr1itEUr
Synthetic	Social; Yu, He, and Liu (2015)	500	-	-

locator (URL). It is noteworthy that datasets like BlogCatalog, Flickr, Twitter, Google+, Facebook, and Orkut are equipped with labelled anomaly types, while TwitterSecurity, EnronInc, and Telecom datasets exhibit temporal characteristics. This classification of datasets based on diverse criteria enhances the understanding and exploration of various anomaly patterns across different application domains.

2.6 APPLICATION DOMAIN OF ANOMALY DETECTION

Anomaly detection, a crucial component in various domains, plays a pivotal role in safeguarding systems against irregularities and malicious activities. This section delves into the diverse application domains of anomaly detection, spanning from network fraud detection to the identification of opinion spam, monitoring financial fraud, information security, and fortifying healthcare systems.

2.6.1 Network Fraud Detection

Network intrusion refers to the illicit use of online technologies to perpetrate fraudulent activities such as creating fraudulent programmes, disseminating false information, and tampering with data (Lin, Liu, Wu, Zuo, Wan, & Li, 2019). These activities involve unlawfully acquiring individuals' private information and assets through deceptive means. Typically, these frauds are based on the daily interactions users engage in as part of their routine lives (Bharne & Bhaladhare, 2023). Common types of fraud include the following:

Hacking fraud: Hacking fraud occurs when hackers use technical means to steal user account passwords (Alsayed & Bilgrami, 2017). Once they gain access to a user's account, they intrude into the social interaction patterns between target users, observe and learn social friends' communication patterns, and send fraudulent financial information (Leonov, Vorobyev, Ezhova, Kotelyanets, Zavalishina, & Morozov, 2021). This often results in financial losses for the target user.

Phishing attack: A phishing attack is defined as attackers exploiting knowledge of a target user's daily habits to engage in fraudulent activities. Deceptive emails and forged websites are used to carry out phishing scams, where victims are coerced into disclosing private information such as credit card numbers, bank account details, and identification numbers (Gupta,

Singhal, & Kapoor, 2016). Fraudsters typically disguise themselves as trustworthy brands, such as online banks, retailers, and credit card companies, to deceive users into divulging their personal information.

Online shopping fraud: In real-life scenarios, users' social media accounts are often linked to e-commerce platforms like Amazon and Alibaba (Weng et al., 2018). Attackers exploit fictitious product and service information or utilise deceptive means, such as misleading links, to defraud consumers of their property.

These fraudulent practices underscore the need for robust security measures and user awareness to safeguard against the ever-evolving landscape of network fraud detection.

2.6.2 Opinion Spam Detection

Opinion spam detection is a critical task in the field of natural language processing and information retrieval, aiming to identify and filter out fake or deceptive reviews and opinions posted on online platforms. The proliferation of opinion spam can significantly impact the credibility of user-generated content, mislead consumers, and distort the overall sentiment analysis of products or services (Rosso & Cagnina, 2017). To address these challenges, researchers and practitioners employ various techniques and methodologies for opinion spam detection.

One common approach involves leveraging machine learning algorithms to analyse textual features within reviews. Features such as sentiment polarity, frequency of specific words or phrases, and writing style can be used to distinguish between genuine and spammy opinions (Ren & Ji, 2019). Supervised learning models, such as SVM or deep learning-based architectures, are often trained on labelled datasets to classify reviews as either authentic or spam (Danilchenko, Segal, & Vilenchik, 2022; Tian, Mirzabagheri, Tirandazi, & Bamakan, 2020).

Another strategy involves considering the contextual information surrounding reviews. Researchers may explore the relationships and interactions between users, the temporal patterns of reviews, and the consensus among multiple reviews for the same product or service (Fahfouh, Riffi, Mahraz, Yahyaouy, & Tairi, 2022). Anomalies in these patterns can be indicative of opinion spam. Furthermore, linguistic and semantic analysis plays a crucial role in opinion spam detection. Natural language processing techniques are employed to identify patterns of deception, such as overly positive or negative sentiment, the use of exaggerated language, or the presence of irrelevant content (Ren & Zhang, 2016).

As the landscape of online reviews continues to evolve, researchers are increasingly exploring advanced methods, including the integration of user behaviour analysis, semi-unsupervised learning approaches, to enhance the accuracy and efficiency of opinion spam detection systems (Ye, Kumar, & Akoglu, 2016; Li et al. 2023). To ensure the reliability of online user opinions for consumers and businesses alike, the development of robust and adaptive models is essential to stay ahead of evolving spamming techniques.

2.6.3 Financial Fraud Monitoring

Financial fraud monitoring in the realm of social networks has become imperative, encompassing various domains such as credit card transactions, Bitcoin exchanges, and banking phishing attempts. The multifaceted nature of financial transactions within social networks demands a nuanced approach to anomaly detection for safeguarding against diverse forms of fraudulent activities (Aburrous, Hossain, Dahal, & Thabtah, 2010).

Credit card transaction fraud: One prominent area of concern is credit card transaction fraud, where malicious attackers exploit vulnerabilities in payment systems. Advanced anomaly detection techniques scrutinise transactional behaviours, patterns, and contextual information to identify irregularities indicative of potential credit card fraud (Dal Pozzolo, Boracchi, Caelen, Alippi, & Bontempi, 2017). Machine learning models, including neural networks, gaining and extracting intricate features from credit card transaction data, enhance the precision of fraud detection systems (Shehnepoor, Togneri, Liu, & Bennamoun, 2021). As shown in Figures 2.4(a) and 2.4(b), this is a classic credit card fraud detection project on Kaggle.[2] This project comprises 284,807 transaction records, each with 28 features, and includes 492 instances of fraudulent behaviour. Figures 2.4(b) and 2.4(b) depict the number of samples labelled and correctly labelled for predictions using logistic regression and SVM models, respectively.

Bitcoin transaction fraud: The rise of cryptocurrency, particularly Bitcoin, has introduced new challenges in fraud detection within social networks. Anomaly detection methods adapted for Bitcoin transactions focus on scrutinising transactional volumes, patterns, and network structures unique to cryptocurrency exchanges (Monamo, Marivate, & Twala, 2016;

[2] https://www.kaggle.com/datasets/mlg-ulb/creditcardfraud

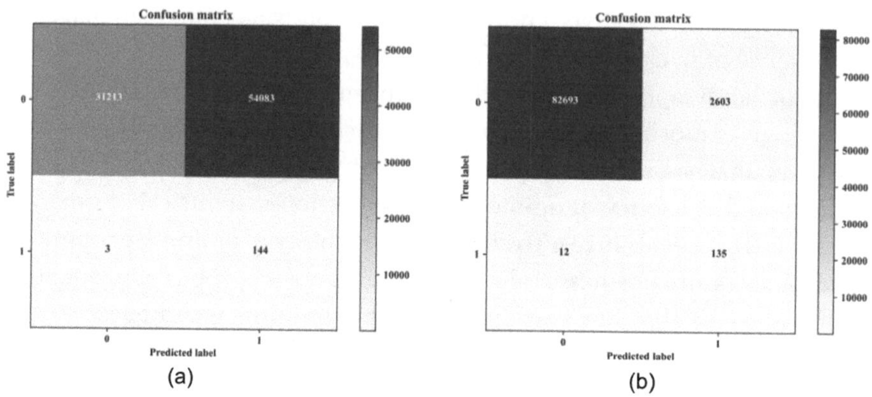

FIGURE 2.4 Credit card fraud detection by logistic regression (a) and SVM (b). Credit card fraud detection plays a crucial role in issuing bank credit cards. Here, we use logistic regression and SVM models to detect anomalies in a Kaggle dataset. These two methods yield different accuracy rates for detecting fraudulent credit cards.

Priya & Saradha, 2021). These methods contribute to the early identification of suspicious activities within the Bitcoin network.

Banking phishing attempts: Banking phishing remains a persistent threat, wherein attackers deploy deceptive emails and forged websites to trick users into revealing sensitive information such as bank account details. Anomaly detection models integrate user profiles, interaction histories, and network structures, and aid in identifying patterns indicative of phishing attempts (Alsayed & Bilgrami, 2017). Machine learning-based approaches, particularly ensemble methods, enhance the ability to discern anomalous patterns in banking phishing scenarios.

2.6.4 Information Security Detection

With the rise of machine learning and deep learning technologies, user posts and online social interactions have become valuable sources for accurately inferring various user attributes, including roles, gender, race, age, political interests, and location. Ensuring information security in online social networks has garnered significant attention in related research, primarily falling into two categories.

(1) Privacy violation detection: The increasing participation of users in location-based social networks (LBSN) for services such as friend finding,

location-based searches, check-ins, and geotagged photo sharing has raised concerns about privacy inference (Caliskan Islam, Walsh, & Greenstadt, 2014). Location information not only reveals an individual's geographical location but also discloses lifestyle, habits, and personal information, exposing users to higher privacy risks. There are some ways to address these risks: (a) personal attribute inference. Malicious activities, such as phishing and identity verification based on personal attributes, are mitigated using methods like *k*-anonymity techniques (Kiabod, Dehkordi, & Barekatain, 2019), game-theoretic defense mechanisms (Manshaei, Zhu, Alpcan, Bacşar, & Hubaux, 2013), and quantifying probability mapping (QPM) (Chen, Lijffijt, & De Bie, 2018) to encrypt and secure users' sensitive information. (b) Contextual inference: Leveraging user-engaged content, topics, and posts, personal private information is inferred, establishing social links and circles among users. Common models include topic bipartite graph embedding (Liao, He, Zhang, & Chua, 2018) and defense methods involving differential privacy (Huang, Zhang, Xiao, Wang, Gu, & Wang, 2020).

(2) Fake news and disinformation detection: Fake news and disinformation detection in social networks represent critical endeavours aimed at identifying and mitigating the spread of deceptive information within online communities. This multifaceted task involves leveraging a combination of natural language processing, network analysis, and machine learning techniques to distinguish credible information from misleading narratives.

The application of information propagation theories in social networks plays a crucial role in sentiment analysis and bias prediction (Pozzi, Fersini, Messina, & Liu, 2016). Analysing the propagation patterns of information within the social network helps uncover anomalies in the spread of fake news. Identifying rapid and widespread dissemination, especially from unverified sources, serves as a key indicator of potential disinformation. On the other hand, evaluating the structure of the social network itself aids in understanding the dynamics of information flow (Shi, Li, Zhang, Sun, & Philip, 2016). Detection methods consider the connections between users, identifying clusters or patterns that may indicate coordinated efforts to spread disinformation (Shu et al., 2020).

2.6.5 Healthcare Insurance Fraud Detection

Health insurance serves as a safeguard for individuals' life and health. For instance, *ActiveHealth*[3] collects users' health data to facilitate health management. Fraudsters exploit the system by submitting numerous fraudulent claims within a short period, disappearing once payments are received. This makes anomaly detection in health insurance systems particularly crucial to mitigate economic losses.

The integration of healthcare insurance fraud detection with social network analysis represents an advanced paradigm designed to fortify the healthcare sector against fraudulent activities. This involves amalgamating methodologies from insurance fraud detection with the intricate dynamics of social networks, aiming to improve precision and efficiency in identifying fraudulent claims within the healthcare insurance domain. The main categories include: (1) social network-based entity profiling—leveraging social network analysis to profile entities within the healthcare ecosystem, including providers, patients, and other stakeholders; analysing relationships reveals abnormal collaboration or referral patterns indicative of fraudulent behaviour (Abdallah, Maarof, & Zainal, 2016). (2) Behavioural analytics in social networks—employing behavioural analytics within social networks to identify deviations from typical interactions among healthcare entities. Anomalous billing practices, unusual patient referrals, or abnormal claim submission patterns trigger alerts for further investigation (Dua & Bais, 2014). (3) Communication pattern analysis—monitoring communication patterns among healthcare entities to reveal irregularities (Abdallah, Maarof, & Zainal, 2016). Instances where providers excessively communicate or collude in an abnormal manner may signal potential fraudulent activities.

By harnessing the power of social network analysis, these integrated approaches contribute to creating a more resilient and secure environment for healthcare insurers, ultimately safeguarding the integrity of insurance claims and improving the overall efficiency of the healthcare system.

---

[3] http://www.activehealth.com

2.7 ONGOING CHALLENGES AND OPPORTUNITIES

Despite the considerable and outstanding efforts in anomaly detection within online social networks, the ongoing development of intelligent social networks and machine learning presents persistent challenges in terms of input data, distributed collaboration across various social media platforms, interpretable anomaly detection, and resilience against adversarial attacks.

2.7.1 Input Data

In the realm of online social networks, user interactions span across various data types, such as text, ratings, videos, images, and links. It is crucial to effectively analyse the input in online social networks and choose corresponding mining techniques. From the perspective of input data, the primary challenges include the following:

Data attributes: The input typically comprises a set of data instances, also referred to as objects, records, points, vectors, events, cases, samples, and entities. Each data instance can be described using a set of attributes (i.e., variables, features, fields, and dimensions). Attributes may take different types, such as binary, categorical, or continuous. Each data instance may contain a single attribute (i.e., univariate) or multiple attributes (i.e., multivariate). In the case of multivariate data instances, all attributes may belong to the same type or be a mix of different data types. On one hand, it is often assumed that data instances are independent of each other. On the other hand, data instances may be interrelated, such as in sequential, temporal, and spatial data, exhibiting specific organisation and strong correlations.

Heterogeneous data: Input data comes in various formats, including text, binary/decimal, videos, and images. Each data format has a unique arrangement, making the extraction of effective features challenging. Transforming diverse features from multiple data formats into a unified feature space remains a formidable problem. Existing efforts often focus on a single data model, neglecting the intricacies of multimodal data fusion (Qi, Zhang, Dou, & Ni, 2017).

Imbalanced data: Firstly, due to the sensitivity of raw data, issues like missing, duplicate, and erroneous data are common. Identifying anomalies in the data, particularly considering the protection of sensitive information, can be time-consuming and may involve manual rule-making. Secondly, anomalous data constitutes a small portion of the overall data.

The scarcity of labelled data, coupled with the challenge of quickly and effectively training models for accurate classification, poses difficulties.

2.7.2 Explainable and Interpretable Anomaly Detection

In machine learning, interpretability refers to the ability of a model's decisions and predictions to be understood, explained, and communicated to non-expert users, enhancing trust in the model's behaviour. Interpretability strives to facilitate an understanding of how machine learning models acquire knowledge, what insights they glean from data, why they render specific decisions for each input, and whether these decisions can be deemed reliable (Zhang, Tiňo, Leonardis, & Tang, 2021). Within the context of social network anomaly detection, interpretability plays a pivotal role in bridging the gap between model predictions and actionable insights. Transparent models not only aid in grasping the reasoning behind anomaly classifications but also cultivate user trust and informed decision-making (Li, Zhu, & Van Leeuwen, 2023c; Qi et al., 2020).

Explainable machine learning: Traditional interpretable methods in anomaly detection include rule-based models (Rajapaksha, Bergmeir, & Buntine, 2020), decision trees (Bai, Li, Li, Jiang, & Xia, 2019), and rule generation-based approaches. These methods heavily rely on existing rules or expert knowledge and often involve only one type of user behaviour data, struggling to adapt to the complexity of user data and the correlation of multiple data types. Machine learning-based interpretability methods possess the powerful ability to learn underlying patterns and regularities in data, demonstrating stronger generalisation capabilities.

Interpretable machine learning models can be classified into two types: ante-hoc interpretability and post-hoc interpretability (Li, Zhu, & Van Leeuwen, 2023). Ante-hoc interpretability involves using machine learning models with simple structures that inherently possess self-explanatory features. The latter refers to interpreting the results and parameters of a trained model based on the different interpretation goals and objects. Naive Bayes models (Wang, Rudin, Doshi-Velez, Liu, Klampfl, & MacNeille, 2017) and generalised additive models, characterised by overall simulatability and decomposability in single steps, offer good ante-hoc interpretability. Furthermore, techniques like local interpretable model-agnostic explanations (Singh & Anand, 2019) and Shapley Additive exPlanations (Antwarg, Miller, Shapira, & Rokach, 2021) provide post-hoc interpretability, helping to understand the decision rationale of complex models like deep neural networks.

Uniform evaluation metrics: In the field of interpretable anomaly detection, a substantial number of anomaly detection predictions often rely on human cognition, qualitative assessment, and the formulation of hard rules, making it challenging to precisely and flexibly evaluate predictions (Guo, Mu, Xu, Su, Wang, & Xing, 2018). From an ante-hoc perspective, quantifying the inherent interpretability of each step within a model depends on the specific application scenarios and the framework's design. From a post-hoc perspective, interpretation methods explaining the reasons for predictions from machine learning models are not always logically consistent and have certain limitations. Moreover, utilising the RMSE metric to measure the difference between ground-truth labels and model-predicted results is limited to evaluating model accuracy, which cannot be directly applied to features' sensitivity analysis, backpropagation, and interpretation consistency domain (Guo, Mu, Xu, Su, Wang, & Xing, 2018).

2.7.3 Adversarial Attacks

An adversarial attack is the intentional design, construction, or selection of specific input samples to deceive, confuse, or attack the inputs to a machine learning model in order to influence the model's output results. The goal of an adversarial attack is to enable the machine learning model to produce misleading or incorrect predictions on the adversarial examples, while the results on the normal inputs remain as unchanged as possible (Dai et al., 2018). For all users registered on social media, there are services such as retweeting, commenting, and posting available through websites or application programming platforms. Attackers can write software or applications in this way to learn the behavioural patterns of normal accounts and control them to send content automatically. Such misleading content posted by attackers could influence political elections, and racial rivalries, and spread terrorist messages. Therefore, it is necessary to identify such fake behavioural patterns and contents by utilising the principles of adversarial attacks.

Social botnet detection: One prevalent strategy employed by attackers is the creation of a multitude of false accounts known as social bots. These bots are designed to mimic human-like social behaviours and norms, engaging in interactions and communications with both humans and other autonomous entities. One approach treats the detection of social bots as a binary classification problem within a supervised framework. However, when confronted with new training features, the lack of generalisability often leads to a significant decline in precision (Wang, Qiao, Xie, Nie, Zhang, & Liu, 2023). These methods struggle with adapting to evolving characteristics and behaviours exhibited by social bots.

Another approach involves observing the behaviour data of accounts and making judgments based on prior knowledge of anomalous account patterns. However, this method tends to lag behind the evolving patterns of anomalous accounts. Modelling social bot detection as a multi-class classification problem, while promising better generalisability, remains highly challenging.

Adversarial training: Adversarial training involves incorporating adversarial examples into the training data, enabling the model to learn robust representations against such samples during the training process. Adversarial generative networks (GANs) are a classic case of generative models, which leverage a generator and discriminator (Creswell, White, Dumoulin, Arulkumaran, Sengupta, & Bharath, 2018). The generator creates samples, and the discriminator distinguishes between real and generated samples. The existing GAN models applied to the exploration of user behaviour records and features in social media encounter significant challenges that limit their scalability:

(1) Adversarial transfer patterns—adversarial transfer patterns emerge when adversarial samples, generated on one model, are utilised to attack another model (Xie et al., 2019). Limited research has delved into applying this transfer pattern across diverse social media platforms. The exploration of adversarial transfer patterns remains an understudied aspect, particularly in the context of different social media environments.

(2) Adversarial rule generation—the adversarial rule generation approach focuses on crafting adversarial rules with the intent to deceive models (Zhou, Yao, Zhao, & Zhang, 2022). By systematically exploring specific rules within the input space, attackers generate adversarial examples capable of misleading the model. The challenge lies in formulating comprehensive adversarial rules that can adapt to the intricate interaction behaviours among users. Addressing this challenge requires exploring and devising a wide range of adversarial

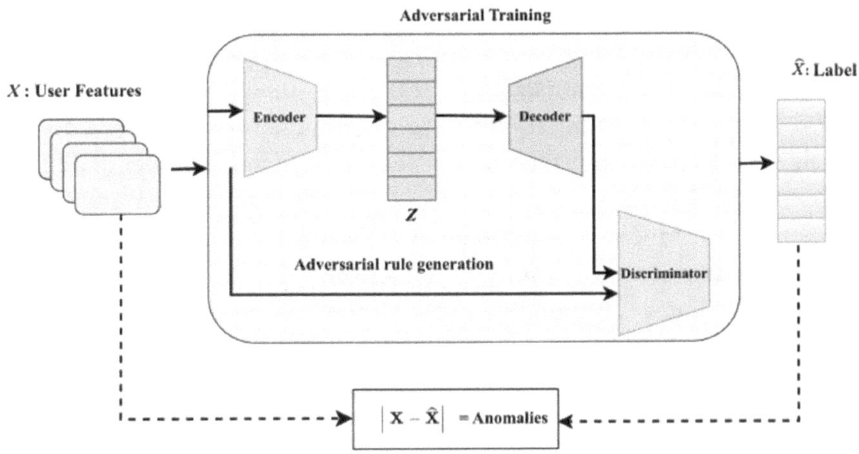

FIGURE 2.5 The promising adversarial network model for predicting anomalies. Generative adversarial networks primarily consist of an encoder and a decoder, and they can be used for anomaly detection through adversarial attacks. By inputting user features into the GAN and leveraging labels, anomalies can be detected effectively.

rules to account for the complexity inherent in user interactions. Specifically, Figure 2.5 illustrates a cutting-edge adversarial network model, which incorporates adversarial training and adversarial rule generation. This model employs an encoder-decoder architecture to represent low-dimensional features for users, and utilises a discriminator to distinguish between normal and anomalous user features.

2.7.4 Distributed Collaborative Anomaly Detection

Cross-platform integration, a pivotal facet of contemporary computing, presents challenges in preserving the integrity and security of interconnected systems (Sedlak, Murturi, Donta, & Dustdar, 2023). This section delves into the critical dimension of anomaly detection within the realm of cross-platform integration, where anomalies, deviations from anticipated patterns or behaviours pose a substantial threat to the smooth operation of distributed platforms.

Multimodal fusion techniques: While individual platforms may be well-suited to specific anomaly detection methods, the fusion of multimodal data demands sophisticated techniques. Integrating textual information with visual data necessitates the development of innovative fusion models such as the Text-Video model Make-a-Video (Singer et al., 2022), the Text-Video detection Clip2video (Fang, Xiong, Xu, & Chen, 2021), and the Speech-Text model Whisper (Cao, Lin, Sun, Lazer, Liu, & Qu, 2012).

Transfer learning for anomaly detection: The utilisation of transfer learning techniques is imperative in the context of cross-platform integration. Models trained on one platform can be fine-tuned to detect anomalies on another, transferring knowledge and adapting to the specific characteristics of each platform (Vincent, Wannes, & Jesse, 2020). Transfer learning methodologies, including pre-trained neural networks and domain adaptation algorithms, contribute to enhancing anomaly detection performance (Chen et al., 2020).

Contextual anomaly detection: Understanding anomalies in the context of cross-platform interactions is crucial. Context-aware anomaly detection models, incorporating temporal and spatial dependencies, enhance the ability to discern anomalies within dynamic integration scenarios (Stojanovic, Dinic, & Stojanovic, 2017; Corizzo, Ceci, Pio, Mignone, & Japkowicz, 2021). Contextual information would allow interpreting the cause of anomalies and improve decision-making (Sedlak, Pujol, Donta, & Dustdar, 2023b).

2.8 CONCLUSION AND DISCUSSION

Social media has captivated millions of individuals, who regularly update and share their daily routines. Exploiting these user-generated content, some malevolent entities use computer technology to mine users' behavioural patterns, engaging in activities such as terrorism and promoting opposing views, thereby negatively impacting people's lives. Consequently, there is an imperative for the investigation and detection of abnormal and malevolent activities within the realm of social media.

In this chapter, we elucidate the definitions of social networks and the anomaly detection domain. Harnessing the foundational knowledge from these two domains, we systematically categorise and classify existing anomaly detection methodologies based on static/dynamic behaviour and static/dynamic structure. Taking a technological development perspective, we draw comparisons between traditional techniques and machine learning-based methods. Additionally, we furnish commonly used datasets along with their corresponding attribute information, contributing to the validation of the rationale behind specific method designs. Various application domains of anomaly detection in social media, including network fraud and financial fraud monitoring, underscore the paramount importance of anomaly detection in this context.

In delving into the persisting challenges and opportunities within social media anomaly detection, we scrutinise four areas, such as data complexity, the interpretability of anomaly results, adversarial attacks, and distributed collaboration. Our objective is to provide prospective research directions and theoretical cornerstones for the field of anomaly detection in social media.

REFERENCES

1. Abdallah, A., Maarof, M. A., & Zainal, A. (2016). Fraud detection system: A survey. *Journal of Network and Computer Applications*, 68, 90–113.
2. Aburrous, M., Hossain, M. A., Dahal, K., & Thabtah, F. (2010). Intelligent phishing detection system for e-banking using fuzzy data mining. *Expert Systems with Applications*, 37(12), 7913–7921.
3. Alsayed, A., & Bilgrami, A. (2017). E-banking security: Internet hacking, phishing attacks, analysis and prevention of fraudulent activities. *International Journal of Emerging Technology and Advanced Engineering*, 7(1), 109–115.
4. Antwarg, L., Miller, R. M., Shapira, B., & Rokach, L. (2021). Explaining anomalies detected by autoencoders using shapley additive explanations. *Expert Systems with Applications*, 186, 115736.
5. Bai, J., Li, Y., Li, J., Jiang, Y., & Xia, S. (2019). Rectified decision trees: Towards interpretability, compression and empirical soundness. *arXiv preprint arXiv:1903.05965.*
6. Baingana, B., & Giannakis, G. B. (2015). Joint community and anomaly tracking in dynamic networks. *IEEE Transactions on Signal Processing*, 64(8), 2013–2025.

7. Bandara, V., Pezeshki, A., & Anura, P. J. (2010). Modeling spatial and temporal behavior of internet traffic anomalies. In *IEEE Local Computer Network Conference* (pp. 384–391). IEEE.
8. Barabási, A.-L. (2009). Scale-free networks: A decade and beyond. *Science*, 325(5939), 412–413.
9. Bharne, S., & Bhaladhare, P. (2023). Comprehensive analysis of online social network frauds check for updates. *Advances in Data-Driven Computing and Intelligent Systems: Selected Papers from ADCIS 2022, Volume 1*, 698, 23.
10. Blaschke, L. M., & Hase, S. (2019). Heutagogy and digital media networks. *Pacific Journal of Technology Enhanced Learning*, 1(1), 1–14.
11. Caliskan Islam, A., Walsh, J., & Greenstadt, R. (2014). Privacy detective: Detecting private information and collective privacy behavior in a large social network. In *Proceedings of the 13th Workshop on Privacy in the Electronic Society* (pp. 35–46). ACM.
12. Cao, N., Lin, Y.-R., Sun, X., Lazer, D., Liu, S., & Qu, H. (2012). Whisper: Tracing the spatiotemporal process of information diffusion in real time. *IEEE Transactions on Visualization and Computer Graphics*, 18(12), 2649–2658.
13. Cao, N., Shi, C., Lin, S., Lu, J., Lin, Y.-R., & Lin, C.-Y. (2015). Targetvue: Visual analysis of anomalous user behaviors in online communication systems. *IEEE Transactions on Visualization and Computer Graphics*, 22(1), 280–289.
14. Chandola, V., Banerjee, A., & Kumar, V. (2009). Anomaly detection: A survey. *ACM Computing Surveys (CSUR)*, 41(3), 1–58.
15. Chaudhary, A., Mittal, H., & Arora, A. (2019). Anomaly detection using graph neural networks. In *2019 International Conference on Machine Learning, Big Data, Cloud and Parallel Computing (COMITCon)* (pp. 346–350). IEEE.
16. Chen, R., Zhang, S., Li, D., Zhang, Y., Guo, F., Meng, W., Pei, D., Zhang, Y., Chen, X., & Liu, Y. (2020). Logtransfer: Cross-system log anomaly detection for software systems with transfer learning. In *2020 IEEE 31st International Symposium on Software Reliability Engineering (ISSRE)* (pp. 37–47). IEEE.
17. Chen, X., Lijffijt, J., & De Bie, T. (2018). Quantifying and minimizing risk of conflict in social networks. In *Proceedings of the 24th ACM*

SIGKDD International Conference on Knowledge Discovery & Data Mining (pp. 1197–1205). IEEE.

18. Corizzo, R., Ceci, M., Pio, G., Mignone, P., & Japkowicz, N. (2021). Spatially-aware autoencoders for detecting contextual anomalies in geo-distributed data. In *International Conference on Discovery Science* (pp. 461–471). Springer.
19. Creswell, A., White, T., Dumoulin, V., Arulkumaran, K., Sengupta, B., & Bharath, A. A. (2018). Generative adversarial networks: An overview. *IEEE Signal Processing Magazine*, 35(1), 53–65.
20. Dai, H., Li, H., Tian, T., Huang, X., Wang, L., Zhu, J., & Song, L. (2018). Adversarial attack on graph structured data. In *International Conference on Machine Learning* (pp. 1115–1124). PMLR.
21. Dal Pozzolo, A., Boracchi, G., Caelen, O., Alippi, C., & Bontempi, G. (2017). Credit card fraud detection: A realistic modeling and a novel learning strategy. *IEEE Transactions on Neural Networks and Learning Systems*, 29(8), 3784–3797.
22. Danilchenko, K., Segal, M., & Vilenchik, D. (2022). Opinion spam detection: A new approach using machine learning and network-based algorithms. In *Proceedings of the International AAAI Conference on Web and Social Media* (Vol. 16, pp. 125–134). Part of the PKP Publishing Services Network.
23. Ding, K., Li, J., & Liu, H. (2019). Interactive anomaly detection on attributed networks. In *Proceedings of the twelfth ACM International Conference on Web Search and Data Mining* (pp. 357–365).
24. Donta, P. K., Sedlak, B., Casamayor Pujol, V., & Dustdar, S. (2023). Governance and sustainability of distributed continuum systems: A big data approach. *Journal of Big Data*, 10(1), 1–31.
25. Donta, P. K., Srirama, S. N., Amgoth, T., & Annavarapu, C. S. R. (2022). Survey on recent advances in IoT application layer protocols and machine learning scope for research directions. *Digital Communications and Networks*, 8(5), 727–744.
26. Dua, P., & Bais, S. (2014). Supervised learning methods for fraud detection in healthcare insurance. *Machine Learning in Healthcare Informatics* (pp. 261–285).
27. Dustdar, S., Pujol, V. C., & Donta, P. K. (2022). On distributed computing continuum systems. *IEEE Transactions on Knowledge and Data Engineering*, 35(4), 4092–4105.

28. Evans, D., Bratton, S., & McKee, J. (2021). *Social media marketing*. AG Printing & Publishing.
29. Fahfouh, A., Riffi, J., Mahraz, M. A., Yahyaouy, A., & Tairi, H. (2022). A contextual relationship model for deceptive opinion spam detection. *IEEE Transactions on Neural Networks and Learning Systems*, 35(1), 1228–1239.
30. Fang, H., Xiong, P., Xu, L., & Chen, Y. (2021). Clip2video: Mastering video-text retrieval via image clip. *arXiv preprint arXiv:2106.11097*.
31. Guare, J. (2016). Six degrees of separation. In *The contemporary monologue: Men* (pp. 89–93). Routledge.
32. Guo, W., Mu, D., Xu, J., Su, P., Wang, G., & Xing, X. (2018). Lemna: Explaining deep learning based security applications. In *proceedings of the 2018 ACM SIGSAC Conference on Computer and Communications Security* (pp. 364–379). ACM.
33. Gupta, S., Singhal, A., & Kapoor, A. (2016). A literature survey on social engineering attacks: Phishing attack. In *2016 International Conference on Computing, Communication and Automation (ICCCA)* (pp. 537–540). IEEE.
34. Huang, H., Zhang, D., Xiao, F., Wang, K., Gu, J., & Wang, R. (2020). Privacy-preserving approach pbcn in social network with differential privacy. *IEEE Transactions on Network and Service Management*, 17(2), 931–945.
35. Jiang, W., Wang, G., Bhuiyan, M. Z. A., & Wu, J. (2016). Understanding graph-based trust evaluation in online social networks: Methodologies and challenges. *ACM Computing Surveys (CSUR)*, 49(1), 1–35.
36. Jiang, Y., & Liu, G. (2022). Two-stage anomaly detection algorithm via dynamic community evolution in temporal graph. *Applied Intelligence*, 52(11), 12222–12240.
37. Khamparia, A., Pande, S., Gupta, D., Khanna, A., & Sangaiah, A. K. (2020). Multi-level framework for anomaly detection in social networking. *Library Hi Tech*, 38(2), 350–366.
38. Kherif, F., & Latypova, A. (2020). Principal component analysis. In *Machine learning* (pp. 209–225). Elsevier.
39. Kiabod, M., Dehkordi, M. N., & Barekatain, B. (2019). TSRAM: A time-saving k-degree anonymization method in social network. *Expert Systems with Applications*, 125, 378–396.

40. Kong, Z., Jones, A., & Belta, C. (2016). Temporal logics for learning and detection of anomalous behavior. *IEEE Transactions on Automatic Control*, 62(3), 1210–1222.
41. Kopp, M., Grill, M., & Kohout, J. (2018). Community-based anomaly detection. In *2018 IEEE International Workshop on Information Forensics and Security (WIFS)* (pp. 1–6). IEEE.
42. Kotler, P., Burton, S., Deans, K., Brown, L., & Armstrong, G. (2015). *Marketing*. Pearson Higher Education AU.
43. Leonov, P. Y., Vorobyev, A. V., Ezhova, A. A., Kotelyanets, O. S., Zavalishina, A. K., & Morozov, N. V. (2021). The main social engineering techniques aimed at hacking information systems. In *2021 Ural Symposium on Biomedical Engineering, Radioelectronics and Information Technology (USBEREIT)* (pp. 0471–0473). IEEE.
44. Li, Y., Ji, Y., Li, S., He, S., Cao, Y., Liu, Y., Liu, H., Li, X., Shi, J., & Yang, Y. (2021). Relevance-aware anomalous users detection in social network via graph neural network. In *2021 International joint conference on neural networks (IJCNN)* (pp. 1–8). IEEE.
45. Li, Y., Wang, X., Zeng, R., Donta, P. K., Murturi, I., Huang, M., & Dustdar, S. (2023). Federated domain generalization: A survey. *arXiv preprint arXiv:2306.01334*.
46. Li, Y., Zhao, Y., Wang, G., Zhu, F., Wu, Y., & Shi, S. (2017). Effective k-vertex connected component detection in large-scale networks. In *Database systems for advanced applications: 22nd International Conference, DASFAA 2017, Suzhou, China, March 27–30, 2017, proceedings, Part II 22* (pp. 404–421). Springer.
47. Li, Y., Zhu, J., Zhang, C., Yang, Y., Zhang, J., Qiao, Y., & Wang, H. (2023). THGNN: An embedding-based model for anomaly detection in dynamic heterogeneous social networks. In *Proceedings of the 32nd ACM International Conference on Information and Knowledge Management* (pp. 1368–1378). ACM.
48. Li, Z., Sun, D., Xu, F., & Li, B. (2017). Social network based anomaly detection of organizational behavior using temporal pattern mining. In *Proceedings of the 2017 IEEE/ACM International Conference on Advances in Social Networks Analysis and Mining 2017* (pp. 1112–1119). ACM.
49. Li, Z., Zhu, Y., & Van Leeuwen, M. (2023). A survey on explainable anomaly detection. *ACM Transactions on Knowledge Discovery from Data*, 18(1), 1–54

50. Liao, L., He, X., Zhang, H., & Chua, T.-S. (2018). Attributed social network embedding. *IEEE Transactions on Knowledge and Data Engineering*, 30(12), 2257–2270.
51. Lin, H., Liu, G., Wu, J., Zuo, Y., Wan, X., & Li, H. (2019). Fraud detection in dynamic interaction network. *IEEE Transactions on Knowledge and Data Engineering*, 32(10), 1936–1950.
52. Luo, X., Wu, J., Beheshti, A., Yang, J., Zhang, X., Wang, Y., & Xue, S. (2022). ComGA: Community-aware attributed graph anomaly detection. In *Proceedings of the fifteenth ACM International Conference on Web Search and Data Mining* (pp. 657–665). ACM.
53. Ma, X., Wu, J., Xue, S., Yang, J., Zhou, C., Sheng, Q. Z., Xiong, H., & Akoglu, L. (2021). A comprehensive survey on graph anomaly detection with deep learning. *IEEE Transactions on Knowledge and Data Engineering*, 35(12), 12012–12038.
54. Manshaei, M. H., Zhu, Q., Alpcan, T., Bacşar, T., & Hubaux, J.-P. (2013). Game theory meets network security and privacy. *ACM Computing Surveys (CSUR)*, 45(3), 1–39.
55. McPherson, M., Smith-Lovin, L., & Cook, J. M. (2001). Birds of a feather: Homophily in social networks. *Annual Review of Sociology*, 27(1), 415–444.
56. Miller, D., Sinanan, J., Wang, X., McDonald, T., Haynes, N., Costa, E., Spyer, J., Venkatraman, S., & Nicolescu, R. (2016). *How the world changed social media*. UCL Press.
57. Molontay, R., & Nagy, M. (2019). Two decades of network science: As seen through the co-authorship network of network scientists. In *Proceedings of the 2019 IEEE/ACM International Conference on Advances in Social Networks Analysis and Mining* (pp. 578–583). ACM.
58. Monamo, P., Marivate, V., & Twala, B. (2016). Unsupervised learning for robust bitcoin fraud detection. In *2016 Information security for South Africa (ISSA)* (pp. 129–134). IEEE.
59. Nawaz, W., Khan, K.-U., & Lee, Y.-K. (2016). A multi-user perspective for personalized email communities. *Expert Systems with Applications*, 54, 265–283.
60. Patro, S., & Sahu, K. K. (2015). Normalization: A preprocessing stage. *arXiv preprint arXiv:1503.06462*.
61. Pozzi, F. A., Fersini, E., Messina, E., & Liu, B. (2016). *Sentiment analysis in social networks*. Morgan Kaufmann.

62. Priya, G. J., & Saradha, S. (2021). Fraud detection and prevention using machine learning algorithms: A review. In *2021 7th International Conference on Electrical Energy Systems (ICEES)* (pp. 564–568). IEEE.
63. Qi, L., Dou, W., & Zhang, X. (2016). Service recommendation based on social balance theory and collaborative filtering. In *Service-oriented computing: 14th International conference, ICSOC 2016, Banff, AB, Canada, October 10–13, 2016, proceedings 14* (pp. 637–645). Springer.
64. Qi, L., Hu, C., Zhang, X., Khosravi, M. R., Sharma, S., Pang, S., & Wang, T. (2020). Privacy-aware data fusion and prediction with spatial-temporal context for smart city industrial environment. *IEEE Transactions on Industrial Informatics*, 17(6), 4159–4167.
65. Qi, L., Wang, R., Hu, C., Li, S., He, Q., & Xu, X. (2019). Time-aware distributed service recommendation with privacy-preservation. *Information Sciences*, 480, 354–364.
66. Qi, L., Zhang, X., Dou, W., & Ni, Q. (2017). A distributed locality-sensitive hashing-based approach for cloud service recommendation from multi-source data. *IEEE Journal on Selected Areas in Communications*, 35(11), 2616–2624.
67. Rajapaksha, D., Bergmeir, C., & Buntine, W. (2020). Lormika: Local rule-based model interpretability with k-optimal associations. *Information Sciences*, 540, 221–241.
68. Ren, Y., & Ji, D. (2019). Learning to detect deceptive opinion spam: A survey. *IEEE Access*, 7, 42934–42945.
69. Ren, Y., & Zhang, Y. (2016). Deceptive opinion spam detection using neural network. In *Proceedings of COLING 2016, the 26th International Conference on Computational Linguistics: Technical papers* (pp. 140–150). ACL Anthology.
70. Rosso, P., & Cagnina, L. C. (2017). Deception detection and opinion spam. In *A practical guide to sentiment analysis* (pp. 155–171). Springer.
71. Ruff, L., Vandermeulen, R., Goernitz, N., Deecke, L., Siddiqui, S. A., Binder, A., Müller, E., & Kloft, M. (2018). Deep one-class classification. In *International Conference on Machine Learning* (pp. 4393–4402). PMLR.
72. Sedlak, B., Murturi, I., Donta, P. K., & Dustdar, S. (2023a). *A privacy enforcing framework for data streams on the edge*. IEEE.

73. Sedlak, B., Pujol, V. C., Donta, P. K., & Dustdar, S. (2023b). Designing reconfigurable intelligent systems with Markov blankets. In *International Conference on Service-oriented Computing* (pp. 42–50). Springer.
74. Shehnepoor, S., Togneri, R., Liu, W., & Bennamoun, M. (2021). Social fraud detection review: Methods, challenges and analysis. *arXiv preprint arXiv:2111.05645.*
75. Shu, K., Bhattacharjee, A., Alatawi, F., Nazer, T. H., Ding, K., Karami, M., and Liu, H. (2020). Combating disinformation in a social media age. *Wiley Interdisciplinary Reviews: Data Mining and Knowledge Discovery, 10(6):e1385.*
76. Singer, U., Polyak, A., Hayes, T., Yin, X., An, J., Zhang, S., Hu, Q., Yang, H., Ashual, O., Gafni, O., et al. (2022). Make-a-video: Text-to-video generation without text-video data. *arXiv preprint arXiv:2209.14792.*
77. Singer, U., Polyak, A., Hayes, T., Yin, X., An, J., Zhang, S., Hu, Q., Yang, H., Ashual, O., Gafni, O., et al. (2022). Make-a-video: Text-to-video generation without text-video data. *arXiv preprint arXiv:2209.14792.*
78. Singh, J., & Anand, A. (2019). EXS: Explainable search using local model agnostic interpretability. In *Proceedings of the Twelfth ACM International Conference on Web Search and Data Mining* (pp. 770–773). ACM.
79. Stojanovic, N., Dinic, M., & Stojanovic, L. (2017). A data-driven approach for multivariate contextualized anomaly detection: Industry use case. In *2017 IEEE International Conference on Big Data (Big Data)* (pp. 1560–1569). IEEE.
80. Tian, Y., Mirzabagheri, M., Tirandazi, P., & Bamakan, S. M. H. (2020). A non-convex semi-supervised approach to opinion spam detection by ramp-one class SVM. *Information Processing & Management*, 57(6), 102381.
81. Vincent, V.,Wannes, M., and Jesse, D. (2020). Transfer learning for anomaly detection through localized and unsupervised instance selection. In *Proceedings of the AAAI Conference on Artificial Intelligence*, volume 34, pages 6054–6061.
82. Viswanath, B., Bashir, M. A., Crovella, M., Guha, S., Gummadi, K. P., Krishnamurthy, B., & Mislove, A. (2014). Towards detecting

anomalous user behavior in online social networks. In *23rd USENIX Security Symposium (USENIX security 14)* (pp. 223–238). USENIX.
83. Vošner, H. B., Bobek, S., Kokol, P., & Krečič, M. J. (2016). Attitudes of active older internet users towards online social networking. *Computers in Human Behavior*, 55, 230–241.
84. Wang, H., & Hu, D. (2005). Comparison of SVM and LS-SVM for regression. In *2005 International Conference on Neural Networks and Brain* (Vol. 1, pp. 279–283). IEEE.
85. Wang, H., Wu, J., Hu, W., & Wu, X. (2019). Detecting and assessing anomalous evolutionary behaviors of nodes in evolving social networks. *ACM Transactions on Knowledge Discovery from Data (TKDD)*, 13(1), 1–24.
86. Wang, L., Qiao, X., Xie, Y., Nie, W., Zhang, Y., & Liu, A. (2023). My brother helps me: Node injection based adversarial attack on social bot detection. In *Proceedings of the 31st ACM International Conference on Multimedia* (pp. 6705–6714). ACM.
87. Wang, T., Rudin, C., Doshi-Velez, F., Liu, Y., Klampfl, E., & MacNeille, P. (2017). A Bayesian framework for learning Rule Sets for interpretable classification. *The Journal of Machine Learning Research*, 18(1), 2357–2393.
88. Weng, H., Li, Z., Ji, S., Chu, C., Lu, H., Du, T., & He, Q. (2018). Online e-commerce fraud: A large-scale detection and analysis. In *2018 IEEE 34th International Conference on Data Engineering (ICDE)* (pp. 1435–1440). IEEE.
89. Xie, C., Zhang, Z., Zhou, Y., Bai, S., Wang, J., Ren, Z., & Yuille, A. L. (2019). Improving transferability of adversarial examples with input diversity. In *Proceedings of the IEEE/CVF Conference on Computer Vision and Pattern Recognition* (pp. 2730–2739). IEEE.
90. Xu, Y., Feng, Z., Xing, M., Wu, H., Chen, S., Xue, X., & Dustdar, S. (2023). Metapath-guided multi-headed attention networks for trust prediction in heterogeneous social networks. *Knowledge-Based Systems*, 282, 111119.
91. Ye, J., Kumar, S., and Akoglu, L. (2016). Temporal opinion spam detection by multivariate indicative signals. In *Proceedings of the International AAAI Conference on Web and Social Media*, volume 10, pages 743–746.

92. Yu, R., He, X., & Liu, Y. (2015). Glad: Group anomaly detection in social media analysis. *ACM Transactions on Knowledge Discovery from Data (TKDD)*, 10(2), 18.
93. Yu, R., Qiu, H., Wen, Z., Lin, C., & Liu, Y. (2016). A survey on social media anomaly detection. *ACM SIGKDD Explorations Newsletter*, 18(1), 1–14.
94. Yu, W., Cheng, W., Aggarwal, C. C., Zhang, K., Chen, H., & Wang, W. (2018). Netwalk: A flexible deep embedding approach for anomaly detection in dynamic networks. In *Proceedings of the 24th ACM SIGKDD International Conference on Knowledge Discovery & Data Mining* (pp. 2672–2681). ACM.
95. Yuan, L.-P., Choo, E., Yu, T., Khalil, I., & Zhu, S. (2021). Time-window based group-behavior supported method for accurate detection of anomalous users. In *2021 51st Annual IEEE/IFIP International Conference on Dependable Systems and Networks (DSN)* (pp. 250–262). IEEE.
96. Zhang, Y., Tiňo, P., Leonardis, A., & Tang, K. (2021). A survey on neural network interpretability. *IEEE Transactions on Emerging Topics in Computational Intelligence*, 5(5), 726–742.
97. Zhou, N., Yao, N., Zhao, J., & Zhang, Y. (2022). Rule-based adversarial sample generation for text classification. *Neural Computing and Applications*, 34(13), 10575–10586.

CHAPTER 3

Anomaly Detection in Healthcare

Lipika Dinda, Govind Narayan Patel, and Jitesh Pradhan

3.1 INTRODUCTION

An anomaly is something that is not consistent with the norm, planned outcome, or accepted standard. It denotes an exception to the norm or a departure from customary behaviour. Scientific, statistical, technological, and general observations are only a few of the domains in which anomalies can be discovered. In scientific contexts, anomalies may point to observations or unexpected results that depart from widely held beliefs or models. In informal conversation (Fernando, Gammulle, Denman, Sridharan, & Fookes, 2021), the expression is frequently employed to describe something notable or uncommon. The details of anomaly detection and its challenges have been described in the following sub-sections.

3.1.1 Anomaly Detection

Finding patterns, behaviours, or data points in a dataset that substantially differ from expected or typical behaviour is known as anomaly detection, a technique used in data analysis and machine learning. The purpose is to find anomalies, or instances in the data that are outliers or diverge from established trends. This approach may help a wide range of sectors, such as banking, healthcare, industrial systems, cybersecurity, and more. An overview of anomaly detection's general operation is given below:

- Data collection: Gather data that demonstrates the usual behaviour or patterns of the system.

DOI: 10.1201/9781003463559-3

- Model training: Utilise the gathered data to train a model or algorithm that will enable it to recognise the typical characteristics and trends of the system.
- Anomaly detection: After training, the model may be used on new or untested data to find and label as anomalies any instances that significantly deviate from the regular patterns that have been developed. Numerous methods and strategies, including statistical methods, machine learning algorithms, and deep learning techniques, are available for anomaly identification.

3.1.2 Common Methods of Anomaly Detection

- Statistical measurements: Utilise statistical measures like mean, median, standard deviation, and clustering techniques to find outliers.
- Machine learning algorithm: Both supervised and unsupervised machine learning algorithms may be put into practice. Unsupervised learning helps uncover undiscovered irregularities since the computer finds regular patterns without labelled input.
- Methodologies for deep learning: Neural networks, and in particular auto encoders, have the ability to interpret complicated patterns in input and identify problems. In many different applications, anomaly detection is crucial for spotting potential issues, fraud, mistakes, or security threats. Proactively addressing problems before they worsen enhances system security and overall reliability for enterprises.

3.1.3 Several of the Main Obstacles in Anomaly Identification

- Labelling data for training: Because supervised anomaly detection relies on labelled data for training, it might be difficult to gather a sufficient number of labelled anomalous samples. As anomalies are often uncommon occurrences, compiling relevant examples for training could be costly and time-consuming.
- Unbalanced datasets: A significant imbalance is frequently seen in datasets used to identify anomalies, with a small fraction of occurrences showing abnormalities and the majority reflecting normal behaviour. This imbalance may cause models to become biased, making it more difficult for them to identify anomalies.

- Dynamic and changing systems: Systems and data distributions can alter over time, creating a dynamic environment. Models may face challenges if they are unable to adapt to variations in usual behaviour or anomalies that were absent from the training set.
- Noise and outliers: It might be challenging to discern between noise or outliers that aren't supported by relevant data and actual irregularities. False positives from noisy data or outliers that might not be true anomalies might impact the anomaly detection system's dependability.
- Interpretability: Certain anomaly detection methods could be difficult to understand, particularly if they rely on intricate machine learning models like neural networks. It becomes crucial to comprehend and communicate the logic behind anomaly detection, particularly for applications that rely on user confidence and understanding.
- Scalability: As datasets get bigger, anomaly detection systems' processing requirements might become a bottleneck. Working with real-time or streaming data raises scalability concerns.
- Adversarial attacks: In particular circumstances, adversaries may intentionally attempt to fool or alter anomaly detection systems. Adversarial attacks can be challenging to protect against, requiring robust models and ongoing vulnerability scanning.
- Contextual understanding: Since anomalies may rely on the environment in which data is produced, it is important to comprehend it. Ignoring contextual information may result in false positive or false negative outcomes.

3.1.4 The Anomaly Detection Issue Facing the Healthcare Sector Has Several Aspects

The task of anomaly detection is complex (Ukil, Bandyoapdhyay, Puri, & Pal, 2016), requiring a deep understanding of many different aspects in order to solve issues and provide workable solutions. The following distinct features of the anomaly detection issue are present:

- Anomaly categories: Three categories exist for anomalies—point anomalies, which are irregularities involving a single data point; contextual anomalies, which depend on the context in which they occur;

and collective anomalies, which are anomalies involving a group of related instances. Determining the specific types of anomalies existing in a given application is a prerequisite to choosing the appropriate detection strategies.

- Data representation: An important part of anomaly detection is feature extraction and data representation. Data preparation for efficient anomaly detection involves feature engineering, dimensionality reduction, and the identification of relevant features.
- Supervised vs. unsupervised learning: It is possible to approach anomaly detection using supervised or unsupervised learning techniques. While unsupervised approaches do not rely on labelled data for anomalies during training, supervised learning entails training models using labelled data that contains both normal and anomalous cases.
- Temporal aspects: Numerous real-world applications rely on time-series data, and anomalies can reveal temporal trends. Finding irregularities across time and considering temporal correlations are important components in applications such as fraud detection, network monitoring, and industrial processes.
- Processing in real time and scalability: Scalable algorithms are essential for anomaly detection, especially when handling massive volumes of data or requiring real-time processing. Using efficient algorithms and parallelisation techniques is necessary for managing high-dimensional or streaming data.
- Domain-specific information: Generally, anomaly identification benefits from domain-specific knowledge. When creating anomaly detection systems, it is critical to understand the surrounding environment, business laws, and customary practices. Important information for feature selection and model evaluation may be obtained from subject matter experts in the field.
- Evaluation metrics: Determining the effectiveness of anomaly detection techniques requires the use of appropriate assessment measures. The AUC_PR (area under the precision-recall curve), recall, F1 score, precision, and AUC_ROC (area under the ROC curve) are common metrics.

- Adversarial considerations: In rare circumstances, deliberate attempts to fool or sway anomaly detection systems may be made. It is critical to consider adversarial attacks and implement protocols to enhance anomaly detection models' resistance.
- Interpretability: The interpretability of anomaly detection models is crucial in situations where human comprehension and confidence are paramount. The ability to explain why a certain episode was identified as odd is necessary for making judgments.
- Dynamic contexts: Anomalies in dynamic situations can arise or vanish over time. Adaptive models that can adjust to changing patterns and variances in data distributions are essential in these sorts of scenarios. Working anomaly detection systems need to take these aspects into consideration and to tailor techniques to specific application requirements. To ensure that the system is reliable and adaptable to changing conditions, data preparation, algorithm selection, and ongoing monitoring and improvement are often integrated.

3.2 TYPE OF ANOMALIES

The intended abnormality type is a critical component of an anomaly detection strategy (Fernando, Gammulle, Denman, Sridharan, & Fookes, 2021; Yang, Qi, & Zhou, 2023). Three primary categories can be used to classify anomalies as shown in Figure 3.1:

3.2.1 Point Anomalies

A data point is considered anomalous (Fernando, Gammulle, Denman, Sridharan, & Fookes, 2021; Yang, Qi, & Zhou, 2023) if its differences from the rest of the dataset are substantial. The majority of research on anomaly identification focuses on this kind of anomaly since it is the most fundamental. Another example from the real world would be to be vigilant about security flaws on a computer network. Network traffic patterns, which may include requests per second, make up the dataset in this instance. For the sake of simplicity, we will focus on only one aspect: the volume of incoming requests from servers. When the server experiences a sudden, sharp spike in requests that significantly deviates from its usual traffic pattern, this would be classified as a point anomaly. A cyber danger, such as a distributed denial

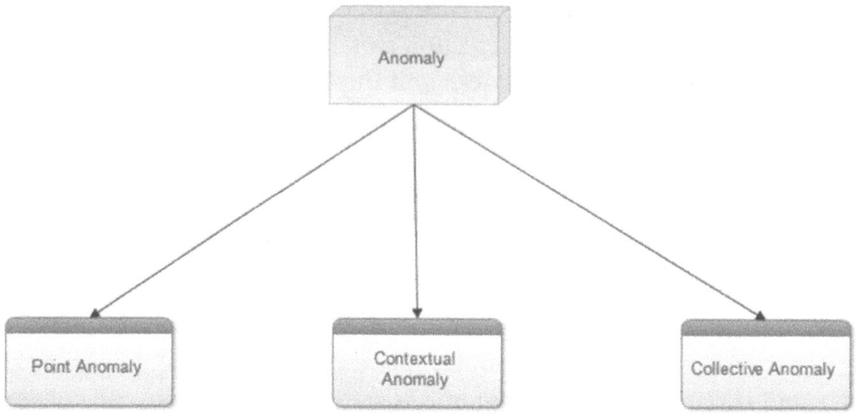

FIGURE 3.1 Categories of anomaly.

of service (DDoS) attack, may be indicated by an occurrence similar Figure 3.2.

3.2.2 Contextual Anomaly

A contextual anomaly is a data instance that significantly deviates from conventional patterns, but only under particular conditions or circumstances. Determining the relevant characteristics or features that provide the dataset with its context is the first step in spotting this sort of anomaly. The subsequent stage entails finding anomalies within this specified environment by searching for glaring departures from expectations. When normality is conditional, it is crucial to recognise these anomalies, necessitating a deep comprehension of how anomalies appear in these specific situations. Each data instance is distinguished by two groups of features: behavioural and contextual. Refer to Figure 3.3.

3.2.3 Collective Anomalies

'Collective anomalies' in anomaly detection refer to situations when a group of data points exhibits abnormal behaviour when examined as a whole, even though each data point seems normal when examined independently. When examining geographical or time-series data, this concept is particularly helpful because it emphasises the links between adjacent or sequential data points.

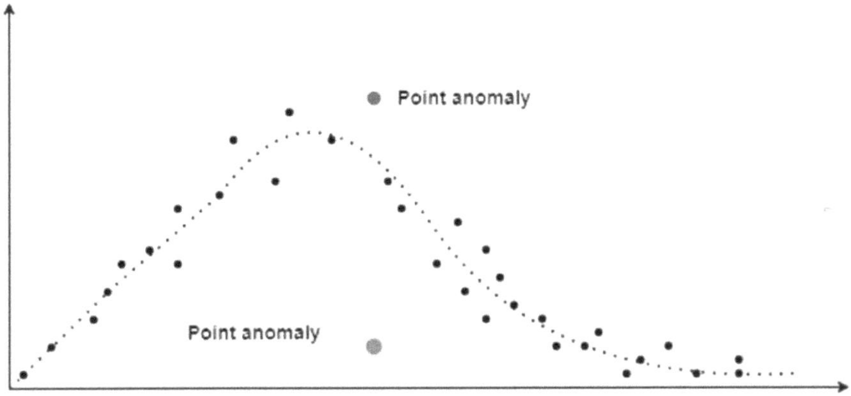

FIGURE 3.2 Point anomalies.

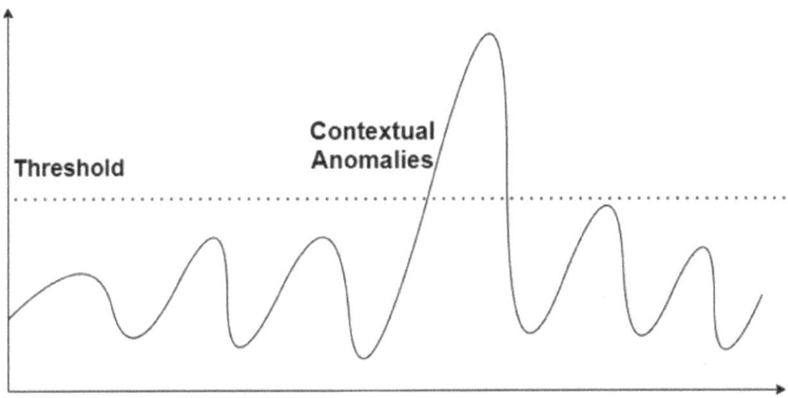

FIGURE 3.3 Contextual anomalies.

For example, in a time-series scenario, a quick and prolonged drop in temperature over a few days may be regarded as a collective anomaly if such a pattern is rare for that time of year, even while individual daily temperature values are not anomalous. On the other hand, when it comes to spatial data, a group of places with a given characteristic—for instance, a concentration of high pollution readings in a small area—might be regarded as a collective anomaly if this pattern deviates from the norm for the region. Finding significant anomalies across a range of industries, including financial market research, network traffic monitoring, environmental monitoring, and healthcare, requires an understanding of the bigger picture

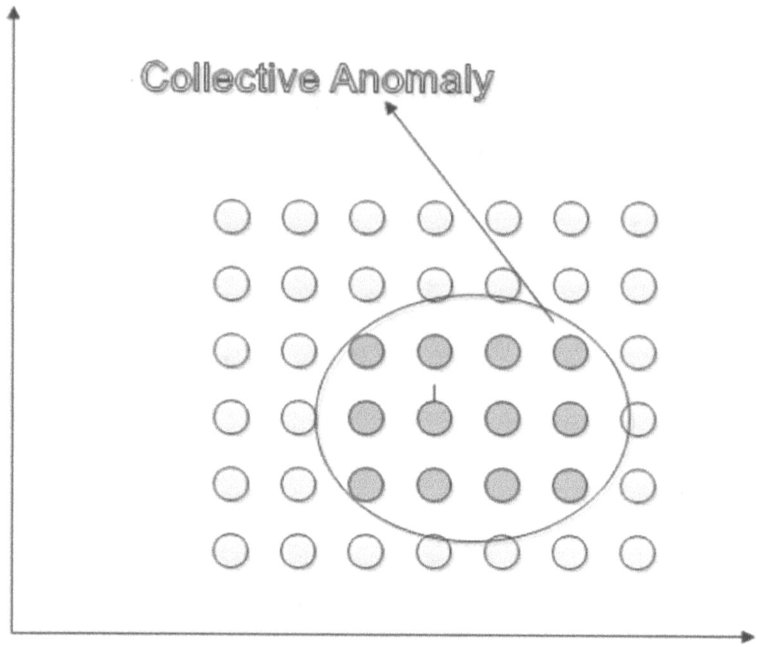

FIGURE 3.4 Collective anomaly.

and the connections between different data points. Collaborative anomaly detection is essential in these domains, and Figure 3.4 shows it.

3.3 APPLICATIONS OF DETECTING ANOMALIES

Several uses for anomaly detection are examined in this section, as summarised in Figure 3.5.

3.3.1 Intrusion Detection

The practice of detecting potentially harmful activity (Chandola, Banerjee, & Kumar, 2009), such as illegal access and computer misuse, within a computer-related system is called as an intrusion detection. These behaviours are especially significant when discussing computer security since they depart from the system's usual behaviour. This is why anomaly detection methods are helpful in the intrusion detection domain. Managing the massive volume of data is a serious challenge to anomaly detection in this field. We need to apply computationally efficient ways to manage

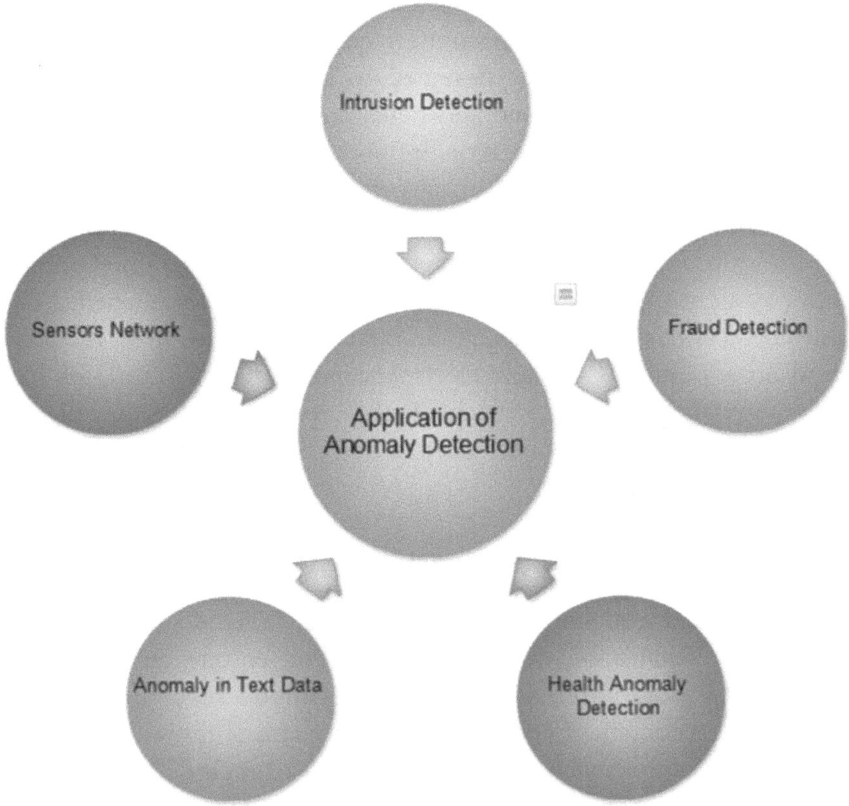

FIGURE 3.5 Application of anomaly detection.

these big datasets. Moreover, the data must often be processed online or in real-time because it is often received in a continuous stream. The amount of incoming data is also a major issue due to the false alarm rate. Given the millions of data pieces involved in the analysis process, an analyst might get overwhelmed by even a small percentage of false alarms. In intrusion detection systems, data is often tagged based on typical behaviour; however, labels for intrusions are sometimes absent. As a result, unsupervised and semi-supervised anomaly detection techniques are preferred in this field.

3.3.2 Medical and Public Health Anomaly Detection

In the medical and public health domains (Chandola, Banerjee, & Kumar, 2009), anomaly identification usually entails reviewing patient records.

Anomalies might result from irregular patient circumstances, mistakes made with instruments, or inaccurate recording of information. A special emphasis is on methods designed to detect disease epidemics in specific regions. As a result, anomaly detection is a crucial problem in this field that requires extreme precision. In this context, patient records might include a variety of data, including age, weight, and blood type, as well as temporal and geographical factors. Most existing anomaly detection methods in this domain try to identify aberrant data, also called point anomalies. Many algorithms use a semi-supervised approach since labelled data mostly reflects healthy patients. Additionally, time-series data, including electrocardiograms (ECGs) and electroencephalograms (EEGs), are handled by anomaly detection approaches in this area. Techniques for collective anomaly identification have been used to find anomalies in this time-series data.

3.3.3 Anomaly Detection within Textual Data

Within this field, anomaly detection methods (Chandola, Banerjee, & Kumar, 2009; Chalapathy & Chawla, 2019) concentrate on pinpointing fresh subjects, occurrences, or news narratives amid a set of documents or news articles. Anomalies in this scenario emerge from innovative and captivating events or unusual topics. The data in this realm is defined as having high dimensionality and sparsity. Moreover, it holds a temporal dimension, given that documents accumulate over time. A notable obstacle for anomaly detection methods in this area involves adeptly managing the substantial variations found in documents associated with a single category or topic.

3.3.4 Sensor Networks

Sensor networks have garnered significant attention in research lately, especially in relation to data analysis. These networks' data, which is gathered from a variety of wireless sensors, has unique properties. Abnormalities in this data might indicate that a sensor is broken or that important events, like invasions, have been discovered. As a result, anomaly detection in sensor networks fulfills a dual function by addressing intrusion detection and sensor defect detection. Sensors collecting many kinds of data, such as binary, discrete, continuous, audio, video, and more, make up a typical sensor network. The deployment environment and the communication channel often create noise and missing values, and the data is generated in a streaming style. Sensor network anomaly detection has special challenges: methods have to be lightweight, run online, and use a distributed data mining methodology for analysis because the data is acquired in a distributed fashion. Sensor data noise complicates anomaly detection by making it harder to distinguish between meaningful abnormalities and unwanted noise or missing values.

3.3.5 Fraud Detection

Fraud detection is the process of locating illegal activities in business settings (Chandola, Banerjee, & Kumar, 2009; Chalapathy & Chawla, 2019), such as stock markets, credit card companies, banks, insurance companies, and mobile phone providers. Malicious users might be people impersonating as clients, a practice known as identity theft, or they could be actual consumers of the company. When these users exploit the organisation's resources without authorisation, fraud takes place. It is important for firms to swiftly identify fraudulent actions in order to avert financial damages. The phrase 'activity monitoring' was first used by Fawcett and Provost (1999) to describe a broad strategy for detecting fraud in various areas. In this application, anomaly detection methods usually entail keeping track of each customer's use profile and keeping an eye out for any abnormalities. Anomaly detection is employed across diverse fields, encompassing areas such as speech recognition, recognising new patterns in robot behaviour, monitoring traffic, safeguarding against fraudulent click-throughs, pinpointing faults in web applications, identifying irregularities in biological datasets, revealing anomalies in census data, detecting correlations in criminal activities, uncovering irregularities in customer relationship management (CRM) data, spotting unusual patterns in astronomical observations, and detecting disruptions in ecosystems.

3.4 ANOMALIES IN HEALTHCARE

Abnormalities in the healthcare sector can manifest in a variety of ways. The instances are shown in Figure 3.6:

- Tracking the health of patients

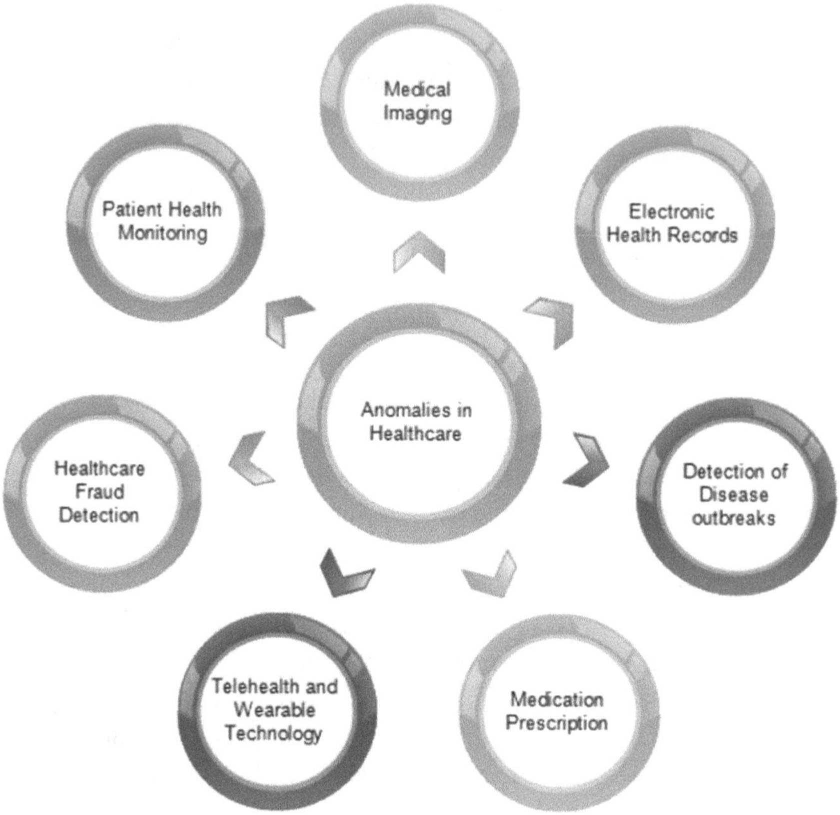

FIGURE 3.6 Anomalies in healthcare.

- *Vital sign abnormalities*: such as high blood pressure, heart rate, or breathing rate, might point to underlying medical issues.
- *Physiological data*: When it relates to physiological data, such as ECG or EEG measurements, anomalies can be identified by departures from typical patterns.

- Medical imaging
  - *Radiology images*: Disturbances from routine medical images, such MRIs or X-rays, may point to abnormalities like tumors, fractures, or other conditions.
  - *Pathological slides*: Vibrations on microscopic inspection slides may indicate abnormalities in the structure of the cells.

- Electronic health records (EHR)
  - *Patient records*: Differencing trends or anomalies in medical history are examples of how inaccuracies or potential health problems may be identified by patient records.
  - *Inconsistencies* in billing and coding data might indicate fraud or errors in the processes involved in medical billing.
- Identification of disease outbreaks
  - *Epidemiological data*: finding deviations from the usual patterns of illness onset and dissemination can help in the early detection of outbreaks.
- Medication prescription and administration
  - *Health information*: Finding anomalies in prescription patterns or medication administration data can help prevent errors and ensure patient safety.
- Wearable technology and telehealth
  - *Data monitoring from a distance*: Anomalies in data obtained from telehealth monitoring systems or wearable technologies may indicate changes in a patient's health status.
- Healthcare fraud identification
  - By *analyzing billing trends* and anomalies in billing data, fraud in the medical billing process may be detected.

3.5 HEALTHCARE DATASETS FOR THE IDENTIFICATION OF ABNORMALITIES

Healthcare-related datasets can be used for anomaly detection research. Several examples of these datasets are as follows:

- PhysioNet: Initially a range of physiological signal datasets, such as those covering vital signs, ECG, and EEG, are available from PhysioNet. These files are perfect for spotting anomalies in patient data.
- Medical Information Mart for Intensive Care III (MIMIC-III): A significant dataset consisting of de-identified medical records from patients in critical care units is provided by MIMIC-III. This dataset,

which includes test results, medication information, and physiological signals, is suitable for examining irregularities in critical care settings.

- Monitoring, epidemiology, and end outcomes, or SEER cancer statistics: By using the datasets on cancer information that SEER offers, researchers may look into anomalies in measures like cancer incidence and survival rates.
- FDA Adverse Event Reporting System (FAERS): FAERS contains reports on adverse events related to drugs and medical devices. It is easy to identify strange trends in adverse event reports by looking for anomalies in this dataset.
- HCUP (Healthcare Cost and Utilization Project): The HCUP provides information on hospital stays, including treatment data, diagnostic codes, and patient profiles. The anomaly identification function of this dataset makes it easier to identify unusual trends in healthcare utilisation.
- Medical Expenditure Panel Survey (MEPS): Offers information on healthcare utilisation, prices, and insurance coverage. Analyzing MEPS data anomalies facilitates the discovery of anomalous spending trends and patterns of healthcare consumption.
- Diabetes dataset: When it comes to diabetic therapy, datasets related to diabetes management—such as information on insulin use, blood glucose levels, and patient demographics—can be utilised to identify anomalies.
- Chest X-ray pictures (pneumonia detection): Using datasets like the chest X-ray dataset for pneumonia diagnosis can assist medical imaging anomaly detection a lot. It aids in the identification of peculiar patterns that indicate diseases such as pneumonia.

3.6 TYPES OF DATA

Generally speaking, there are three main categories of biomedical signals: electrical biomedical datasets, biomedical images, and other sorts of biomedical data. (Fernando, Gammulle, Denman, Sridharan, & Fookes, 2021; Šabić, Keeley, Henderson, & Nannemann, 2021). This third group encompasses a broad range of data kinds, including information gleaned from wearable medical equipment, audio recordings, and laboratory test results.

3.6.1 Radiography (X-Ray)

X-rays, with their shorter wavelengths compared to visible light, have the capacity to pass through most kinds of human body tissues. However, due to the higher density of calcium in bones, X-rays tend to scatter upon contact. A film placed on the side opposite to the X-ray source captures this interaction as a negative image, where areas receiving more X-ray exposure appear darker. Thus, as X-rays easily pass through tissues such as muscles and lungs, these appear darker on the film, whereas bones, which scatter the X-rays, show up as brighter areas. X-ray imaging, widely used for various diagnostic purposes, identifies bone fractures, dental issues, pneumonia, and certain tumors. This method makes use of the various X-ray absorption levels of various tissues to provide important diagnostic data.

3.6.2 Computed Tomography (CT) Scan

A fine X-ray beam is used in CT (computed tomography) imaging to create cross-sectional body scans as the subject is quickly rotated. In order to create a three-dimensional picture, this technique entails obtaining many cross-sectional slices. More fine details are available in this 3D depiction than in traditional X-ray pictures. CT scans are widely used as a diagnostic technique to find injuries or illnesses in different body areas. They are very helpful in localising brain traumas, tumors, blood clots, and tumors or lesions in the belly. Furthermore, because CT scans offer a thorough and complete image of the inside structures of the body, they are essential for the diagnosis of complicated bone fractures and bone malignancies.

3.6.3 MRI (Magnetic Resonance Imaging)

By influencing proton alignment inside the body, a magnetic field is used in magnetic resonance imaging (MRI) to produce pictures. Protons in the human body normally rotate, producing a tiny magnetic field. These protons align themselves with the strong magnetic field produced by an MRI equipment. This alignment is then momentarily disrupted by a radio frequency pulse. The protons release energy when the pulse ends and they align with the magnetic field again. Because different tissues release varying amounts of energy, an MRI scan may distinguish between different body parts.

3.6.4 Electrocardiogram (ECG)

ECG (electrocardiogram) is a tool utilised to visualise the electrical activity flowing through the heart, which is responsible for creating the heartbeat and initiates from the upper part of the heart, progressing to the lower part. When the heart is at rest, heart cells carry a negative charge compared to the external environment. However, during depolarisation, these cells shift to a positive charge. The ECG captures this variation in polarisation. Analyzing the ECG allows for the extraction of two kinds of information. Initially, by assessing time intervals on the ECG, one can identify irregular electrical activities in the heart. Secondly, the magnitude of the electrical activity offers insights into areas of the heart that might be strained or under stress. ECG serves as a valuable diagnostic tool for assessing the electrical health of the heart and detecting potential abnormalities.

3.6.5 Electroencephalogram (EEG)

An electroencephalogram, or EEG, is used to identify the brain's electrical activities that promotes electrical impulse transmission. Tiny electrodes or metal discs, are applied to the scalp in preparation for this treatment. These electrodes record electrical impulses from the brain and subsequently enhance the signals to better brain activity imaging. EEG is a helpful tool for monitoring and assessing brain activity that can aid in the diagnosis and understanding of a variety of neurological conditions.

3.6.6 Magnetoencephalography (MEG)

An electroencephalogram records the electrical fields produced by extracellular currents in the human brain, whereas magnetoencephalography primarily identifies the magnetic fields induced by these currents. It is noteworthy to emphasise that while MEG is not only classified as an electrical biomedical signal, we include it in this category, as it records a derivative of the electrical activity in the brain. The ability of MEG signals to detect abnormal brain activity and circumstances has been the subject of several studies.

3.7 WHY ARE MEDICAL ANOMALIES DIFFERENT?

Medical anomalies differ significantly from other types of aberrations due to the complexity and dynamic nature of biological systems. There are several reasons to emphasise the distinctiveness of medical abnormalities:

- The complexity of biology: The complexity and diversity of human anatomy and physiology is one of the main obstacles. Biological systems are intricate webs of interrelated activities, making it difficult to provide a standard reference point for what is considered 'normal.'
- Individual unpredictability: Since each individual's body is unique, what is normal for one person might be abnormal for another. Variability in lifestyle choices, genetic makeup, and environmental factors all contribute to notable variations in health profiles.
- Dynamic health dynamics: The human body is always evolving, resulting in dynamic health dynamics. Time-related factors such as aging, hormone fluctuations, and environmental impacts add variability to health indices.
- Interpretation of subjective symptoms: Because individuals perceive and express their experiences differently, the symptoms that patients describe are subjective in nature. It becomes much more challenging to diagnose anomalies based only on reported symptoms because of this subjectivity.
- Multiple origins: Medical disorders often result from a combination of environmental, genetic, and behavioural influences. Anomalies are challenging to attribute to a single source due to the intricate interplay of several variables.
- Diagnostic complexity: It may be challenging to quickly discover abnormalities due to the varying accessibility and accuracy requirements of diagnostic instruments for various medical disorders. The identification of health issues could be delayed if certain deviations are not immediately apparent.
- Progressive medical insights: Diagnostic standards and medical understanding are always changing. With time, medical knowledge may lead to a redefined or improved classification of what was formerly thought to be an oddity.

- Impact of contextual factors: A patient's medical history, coexisting conditions, and unique situation all have a significant role in the context of an abnormality and how essential it is understood. The process of detecting anomalies in healthcare is made more challenging by contextual considerations.

3.8 USING MACHINE LEARNING METHODS TO IDENTIFY ANOMALIES

By using real behavioural data as a reference (Kavitha et al., 2021), machine learning, and more especially ML approaches, are widely utilised to find deviations from typical behaviour in system data. The goal is to identify any behaviour that differs from the accepted norm and classify it as abnormal. Three primary categories comprise machine learning techniques: reinforcement learning, unsupervised learning, and supervised learning. The kind of dataset determines which categorisation method is used. While unsupervised approaches are better at classifying and making predictions from unlabelled data, supervised learning works well with labelled data (Tumuluru, Lakshmi, Sahaja, & Prazna, 2019; Chalapathy & Chawla, 2019; Chen & Hengjinda, 2021). ML methods are essential for making decisions in the healthcare industry. Datasets with abnormal patterns are often less prevalent than ones with regular patterns. The k-means algorithm is recommended as an unsupervised method for identifying and categorising aberrant patterns. k-means works well when processing unlabelled data. Its basic concept is straightforward: each data point is repeatedly assigned to one of the k clusters according to attribute similarity in order to identify data groupings. The method picks centroids for the clusters by minimising the sum of squares inside each cluster, as illustrated in Equation (1). These centroids are vital for labelling new formations. When processing sensor data, if items deviate from normal behaviour and don't fit within established clusters, the anomaly detection method recognises the abnormalities and alerts relevant users. In the k-medoid clustering techniques, it identifies some points in the cluster with the least dissimilarity to other points in the same cluster. Dissimilarity (E) with other points is calculated using

Equation (2), and the cost (C) in *k*-medoids is determined by Equation (3).

$$\sum_{i=0}^{n} uj = c^{min} \ [xi - uj]^2 \tag{3.1}$$

In Equation (1), *c* refers to cluster, *xi* is sample, and *uj* is its mean.

$$E = |C_i - P_i| \tag{3.2}$$

In Equation (2), *Ci* refers to cluster, and *Pi* refers to an object in that cluster.

$$C = \sum_{C_i} \sum_{P_i \in C_i} E \tag{3.3}$$

In Equation (3), *C* refers to cost.

3.8.1 *k*-Means Algorithm

INPUT: Dataset

Steps of algorithm (Kavitha et al., 2021):

(1) Set the value of K, representing the number of clusters.

(2) Randomly initialise the *k*-means.

(3) Assign every data point to the closest mean.

(4) Calculate the new mean for every cluster.

(5) Iterate through STEP 3 and STEP 4 until the mean value stabilises and no longer changes.

3.8.2 *k*-Medoid Algorithm

Steps of algorithm (Kavitha et al., 2021)

(1) Select *k* random points as medoids from a set of n data points.

(2) Utilise a distance metric to assign each data point to the nearest medoid.

(3) Iteratively perform the following steps as long as the cost decreases. For each medoid (m) and each data object (o) that is not a medoid:

- Swap m and o, associate each data object with its closest medoid, and recalculate the cost.
- If the cost is higher than in the previous step, revert the swap operation.

3.8.3 Performance Analysis

Validation of the model involves assessing its performance through accuracy and runtime metrics. These metrics are determined using two sets of clusters denoted as S = C1, C2, C3, …, Cn, generated by the clustering algorithm, and the true cluster labels on the data represented as P = D1, D2, D3, …, Dm. To evaluate the association between S and P, each pair of items in the data is categorised as either positive or negative in both sets S and P. In this context, a positive categorisation indicates that the pair of items belong to the same cluster, while a negative categorisation implies that the pair of items do not belong to the same cluster. Accuracy refers to the ratio of correct predictions for test data compared to train data, and its representation is illustrated in Equation (4).

$$A = \frac{TP + TN}{TP + TN + FP + FN} \tag{3.4}$$

The primary objective is to recognise abnormal patterns by leveraging insights from normal patterns. The data partitioning mechanism facilitates the classification of these variations in the data. Both *k*-medoids and *k*-means are utilised to categorise the data based on the segmentation process.

3.8.4 Results

Several methodologies as well as datasets are applied in anomaly detection which uses machine learning and deep learning techniques. Some methods are fusion-based ANN, fusion-based MRI, and FDG-PET. Other machine learning methods include datasets such as Video-EEG and ECG, CheXpert, Chest X-ray8, and ABIDE (Autism Brain Imaging Data Exchange). Some of these give more than 80% accuracy, some techniques identify the diseases, and moreover, some methods give 60% accuracy irrespective of previously available data. All the datasets, methods, and their results are given in Table 3.1.

TABLE 3.1 Comparison of machine learning techniques

Data type/method	Dataset name	Task	Result analysis
X-ray	CheXpert	Chest radiography	The collection comprises 224,316 chest radiographs from 65,240 individuals who have been identified as having 14 chest radiographic findings, and accuracy is 79.3% (Irvin et al., 2019).
X-ray	Chest X-ray8	Bone radiography	Contains text that has been mined for eight illness labels from 108,948 front-view X-ray pictures of 32,717 distinct individuals (Wang, Peng, Lu, Lu, Bagheri, & Summers, 2017).
MRI	ADNI	Assessment of Alzheimer's disease	There are 819 people in the dataset: 229 are normal, 398 have moderate cognitive impairment, and 192 have Alzheimer's disease. Accuracy is 67% (Risacher et al., 2013).
MRI	ABIDE (Autism Brain Imaging Data Exchange)	Autism disorder identification	There were 573 healthy control persons and 539 people with autism spectrum disorders with an accuracy 83.2% (Di Martino et al., 2014).
Ultrasound	EchoNet-Dynamic	Cardiac functionality assessment	Contains human expert comments (calculations, tracings, and measurements) on 10,030 labelled echo cardiogram films (Ouyang et al., 2020).
CT	Lung Image Database Consortium (LIDC)	Lung cancer detection	Contains 244,527 photos from 1,018 participants in its data (Di Martino et al., 2014).
CT	RIDER Lung CT	Lung cancer detection	Thirty-two lung cancer patients had two CT scans of their chests separated by 15 minutes with an accuracy 83% (Zhao et al., 2009).
Body temperature and pulse rate	PPG-DaLia	Monitor temperature level and heart rate level	Because body temperature and pulse rate have a linear relationship, persistent increases or decreases in body temperature or unusual spikes and falls in temperature point to an abnormality (Kavitha et al., 2021).
Night time ECG recordings	PhysioNet	Identification of sleep-related breathing disorders (Penzel, Moody, Mark, Goldberger, & Peter, 2000)	By taking apneas and hypopneas together, sensitivity increased to 93.1%.
Video-EEG and ECG	In-house dataset	Cardiac arrhythmia	During pre-ictal period heart rate was higher with an accuracy 74% (Tigaran, Mølgaard, McClelland, Dam, & Jaffe, 2003).
Heart rate variability (HRV)	In-house nominal 24-hour electrocardiographic (ECG) dataset	Automatic prediction of cardiovascular and cerebrovascular events	By applying random forest model, they identify high-risk hypertensive patients with sensitivity and specificity rates of 71.4% and 87.8%, respectively (Melillo et al., 2015).

3.9 DEEP LEARNING TECHNIQUES IN DETECTING ANOMALIES

Because deep learning (Fernando, Gammulle, Denman, Sridharan, & Fookes, 2021) offers an answer to the aforementioned problems, it is becoming more and more popular in the field of biomedical engineering. The ability of deep learning to represent non-linearity is one of its noteworthy characteristics. By making the model more non-linear, discrepancies in the data may be more successfully captured and normal and anomalous samples can be distinguished. Another advantage of deep learning is its ability to learn features automatically. The availability of vast amounts of data and increased computing power have strengthened deep learning's method of hierarchical feature learning, eliminating the requirement for manual anomaly generation and description. Using neural network design, deep learning can also automatically identify data's long-term relationships; this eliminates the need for explicit feature development. This is an intriguing property of deep learning. For example, by utilising so-called 'memory,' recurrent frameworks such as LSTM (long short-term memory) and GRU (gated recurrent units) may effectively replicate temporal correlations in time-series data.

3.9.1 Algorithmic Methods for Detecting Medical Anomalies

In this section, an exploration of variations in dimensionality across diverse data types and an examination of how contemporary deep learning research addresses and handles these distinctions are outlined.

3.9.1.1 Unsupervised Anomaly Detection (UAE)

In the context of UAE, there is an absence of a supervisory signal during training, meaning that there is no guidance indicating whether a sample is considered abnormal or normal. AEs (autoencoders) and GANs (generative adversarial networks) are two prevalent architectures in unsupervised deep learning, and the subsequent sections introduce and discuss them.

Auto encoders (AEs): Auto encoders (Ballard, 1987) serve as a technique for pre-training deep neural networks, extensively employed for the automated learning of features. These models are symmetrical, and their training involves reconstructing the input from a compressed representation acquired at the central point of the architecture.

In a formal context, considering a dataset with N samples, where the present input is denoted as x, f, and g represent the encoder and decoder networks respectively. The compressed representation, z, is defined as follows:

$$[H]\,Z = f(x) \tag{3.5}$$

and reconstructed using

$$[H]\,y = g(z) \tag{3.6}$$

The model undergoes training to minimise the loss associated with reconstruction.

$$\sum_{x \in N} L\,(x, g\,(f(x))) \tag{3.7}$$

where L represents a distance metric, often adopting the form of mean squared error (MSE).

Various versions of autoencoders exist, including the sparse autoencoder (S-AE). The introduction of a sparsity constraint restricts the count of non-zero elements in the encoded representation, denoted as z. This constraint is implemented through an additional penalty term incorporated into the loss function (i.e., Equation (7)).

$$L_S - AE = \sum_{x \in N} L\,(x, g\,(f(x))) + \lambda \frac{1}{|N|} \sum_{x \in N} |f(x)| \tag{3.8}$$

Here, λ serves as a hyperparameter that governs the intensity of the sparsity constraint.

The **variational autoencoder (VAE)** functions under the assumption that the observations, denoted as x, are drawn from a probability distribution. The goal is to estimate the parameters characterising this distribution. While autoencoders have interesting features, their capacity to model high-dimensional data distributions is limited, often leading to inaccurate reconstructions and imprecise approximations of the modelled data distribution. Denoising autoencoders are made to reconstruct an uncontaminated signal from a noisy (corrupted) input. The goal is to apply the denoising capability to obtain robust and widely applicable feature encoding.

Generative adversarial networks (GANs): Another type (Goodfellow et al., 2014) of auto encoders is generative adversarial autoencoders, commonly referred to as GANs. GANs involve the training of two networks, a

Generator (G) and a Discriminator (D), engaged in a minimax game. The objective is for G to deceive D, while D endeavours to avoid being deceived.

3.9.1.2 Supervise Anomaly Detection

Given the need for a heightened level of sensitivity and resilience (Zhang & Sabuncu, 2018), particularly in diagnostic contexts, supervised learning has been extensively employed in medical anomaly detection. It has shown superior performance compared to unsupervised methods. Unlike unsupervised learning approaches, supervised anomaly detection involves the provision of a supervised signal indicating which instances belong to the normal category and which are considered anomalous. Consequently, this becomes a binary classification task, with models typically trained using binary cross-entropy loss.

$$L = -ylog(f(x)) - (1 - y)\, log(1 - f(x)) \tag{3.9}$$

Here, y represents the actual label, f denotes the classifier, and x stands for the model input.

3.9.1.3 Recurrent Neural Network (RNN)

Recurrence is a pivotal feature, particularly in tasks like time-series modelling, signifying that the result of the current time step is used as an input for the subsequent time step. In the context of medical data analysis, this is crucial for handling sequential data such as EEG and phonocardiographic data, where our focus lies in understanding the temporal progression of the signal as shown Figure 3.7.

3.9.2 Results

Several methodologies as well as datasets are applied in anomaly detection in the deep learning technique. Some of methods are fusion-based ANN, fusion-based MRI, and FDG-PET. Some other datasets are CHB-MIT scalp EEG (sEEG) datasets and Freiburg intracranial EEG (iEEG), Dataset of Kaggle (2016), The Glioma dataset (1426), meningioma (708), and pituitary (930). As a result, some of these give more than 80% accuracy, and some give below 60% accuracy. A lists of datasets, methods, and their results are given in Table 3.2.

TABLE 3.2 Comparison of deep learning techniques

Data type/method	Dataset name	Task	Result analysis
Myelin water imaging of MRIs and fusion of T1-weighted MRIsand	In-house data (44 healthy controls and 55 relapse-remitting MS patients)	Diagnosis (Yoo et al., 2018) of multiple sclerosis	87.9% precision
MRI and FDG-PET are fused	Database of Alzheimer's Disease Neuroimaging Initiative (ADNI) (Lu, Popuri, Ding, Balachandar, & Beg, 2018)	Detection of Alzheimer's disease	Accuracy of 82.93%
Apparent diffusion coefficients of MRIs and T2-weighted MRI images are fused	In-house data (450 noncancerous images and 463 cancer lesions)	Detection of prostate cancer lesions	Specificity is 95.73% and sensitivity is 89.85% (Le et al., 2017)
Decision level fusion using ensemble of classifiers (Islam & Zhang, 2018)	OASIS dataset	Classification between four Alzheimer's disease classes (unaffected, very mild, mild, and moderate)	94% accuracy and 93% recall
Auto-encoders using multi-stage training (Zeng et al., 2018)	In-house (474 schizophrenia and 607 healthy controls)	Schizophrenia diagnosis	85% accuracy
Auto-encoders using multi-stage training (Shehata et al., 2018)	In-house data (100 subjects)	Acute renal transplant rejection is detected early	97% accuracy
Neural memory network (Fernando et al., 2020)	The dataset of glioma (1426), meningioma (708), pituitary (930)	Classification of brain tumours	97.52% accuracy
Fine-tuning AlexNet (Fan, Xu, Fan, Wei, & Li, 2018)	In-house data (690 erosion images and 500 ulcer)	Ulcer and erosion detection	Erosion and ulcer detection accuracy scores are 95.16% and 95.34%, respectively
Short-term Fourier transform (STFT) features are trained using 2D CNN architecture (Truong, Nguyen, Kuhlmann, Bonyadi, Yang & Kavehei, 2017)	CHB-MIT scalp EEG (sEEG) datasets and Freiburg intracranial EEG (iEEG)	Seizure prediction	81% sensitivity
Multi-scale CNN (Hussein, Ahmed, Ward, Wang, Kuhlmann, & Guo, 2019)	Dataset of Kaggle (2016)	Seizure prediction	87.85% sensitivity

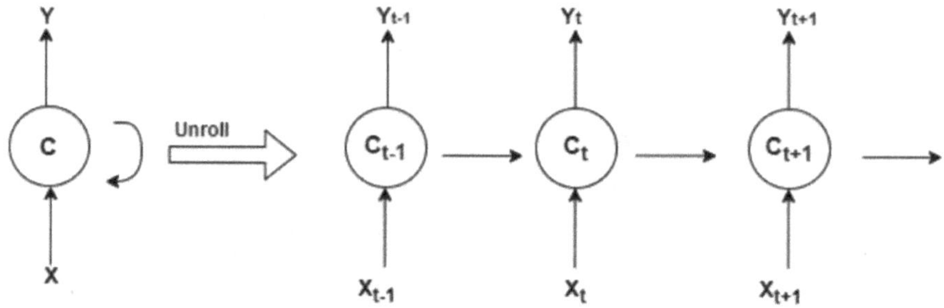

FIGURE 3.7 Recurrent neural network.

3.10 PRINCIPAL CHALLENGES WITH HEALTHCARE DATA

The industry is confronted with several challenges related to the management and utilisation of healthcare data (Fernando, Gammulle, Denman, Sridharan, & Fookes, 2021; Chandola, Banerjee, & Kumar, 2009). These challenges include:

- Information security and privacy: Adhering to HIPAA rules—strict regulations, such as the US's Health Insurance Portability and Accountability Act (HIPAA), make it challenging to preserve compliance while safeguarding patient information.
- Interoperability: Creating standard issues disparate healthcare data sources and systems, non-standard data formats, coding methods, and communication protocols all hinder interoperability by making it harder for information to move freely between various healthcare systems.
- Data integration: Information clusters—the provision of comprehensive patient care and the extraction of important insights are hampered by the inability to combine healthcare data stored in various systems.
- Completeness and accuracy of data: Decision-making and patient care are compromised by insufficient or inaccurate data. While it's important to maintain data quality, problems like data input mistakes and out-of-date information might occur.
- Big data difficulties: Volume, speed, and diversity—the swift increase in healthcare data volume, generating speed, and variety of forms

present technological difficulties for organising and evaluating sizable and heterogeneous information.

- Moral aspects to consider: Informed consent—there are ethical issues in balancing patient privacy and informed consent requirements with the use of patient data for research and healthcare improvement. Careful adherence to ethical norms is necessary to meet these criteria.
- Guidelines and standards for data governance: To prevent problems with data security, quality, and compliance, it is crucial to establish efficient data governance procedures, which include outlining standards, rules, and roles for data management.
- Adoption and integration of technological advancements: Adopting new technologies such as telemedicine, artificial intelligence (AI), and electronic health records (EHRs) requires a large financial outlay in addition to challenges with adoption, user training, and integration.
- Patient engagement: Information accessibility obstacles—technology literacy, data accessibility, and patient engagement are the main obstacles to giving patients access to their health information and involving them in decision-making.
- Cyber security threats and data breaches: Because sensitive data is involved, the healthcare sector is vulnerable to cyber-attacks. Safeguarding healthcare systems against cybersecurity intrusions and guaranteeing the privacy and accuracy of patient information are ongoing concerns. Healthcare providers, legislators, technology suppliers, and other stakeholders must work together to design and implement solutions that prioritise patient care, data security, and regulatory compliance in order to address these difficulties.

3.11 CONCLUSION

Outliers, noise, exceptions, and variations of the system's true behaviour are all considered anomalies. In this chapter, the contributions of many authors to the field of healthcare anomaly detection have been examined. The need for anomaly detection and smart healthcare services has been discussed. For tasks involving prediction and classification, machine learning techniques are extensively employed. Healthcare data is partitioned using the k-means and k-medoid algorithms, and the performance is evaluated in terms of anomaly detection. Additionally, we have discussed a variety of methods for deep learning-based medical anomaly identification. Specifically, we have described several settings for data collection in various medical applications, many deep learning architectures inspired by these diverse data types and issue descriptions, and distinct approaches to learning that have been used. State-of-the-art approaches may now be compared and contrasted despite application disparities, thanks to this systematic study of deep medical anomaly detection research methodology. Furthermore, we presented a thorough synopsis of deep model interpretation tactics, detailing the advantages and disadvantages of such interpretation methods. We discussed the main restrictions on medical anomaly detection methods in our closing remarks.

REFERENCES

1. Ballard, D. H. (1987). Modular learning in neural networks. In *Proceedings of the sixth National Conference on Artificial Intelligence—volume 1* (pp. 279–284). ACM.
2. Chalapathy, R., & Chawla, S. (2019). Deep learning for anomaly detection: A survey. *arXiv preprint arXiv:1901.03407.*
3. Chandola, V., Banerjee, A., & Kumar, V. (2009). Anomaly detection: A survey. *ACM computing surveys (CSUR)*, 41(3), 1–58.
4. Chen, D. J. I. Z., & Hengjinda, P. (2021). Early prediction of coronary artery disease (CAD) by machine learning method—a comparative study. *Journal of Artificial Intelligence and Capsule Networks*, 3(1), 17–33.
5. Di Martino, A., Yan, C.-G., Li, Q., Denio, E., Castellanos, F. X., Alaerts, K., ... others (2014). The autism brain imaging data exchange: Towards a large-scale evaluation of the intrinsic brain architecture in autism. *Molecular Psychiatry*, 19(6), 659–667.
6. Fan, S., Xu, L., Fan, Y., Wei, K., & Li, L. (2018). Computer-aided detection of small intestinal ulcer and erosion in wireless capsule endoscopy images. *Physics in Medicine & Biology*, 63(16), 165001.
7. Fernando, T., Denman, S., Ahmedt-Aristizabal, D., Sridharan, S., Laurens, K. R., Johnston, P., & Fookes, C. (2020). Neural memory plasticity for medical anomaly detection. *Neural Networks*, 127, 67–81.

8. Fernando, T., Gammulle, H., Denman, S., Sridharan, S., & Fookes, C. (2021). Deep learning for medical anomaly detection—a survey. *ACM Computing Surveys (CSUR)*, 54(7), 1–37.
9. Goodfellow, I., Pouget-Abadie, J., Mirza, M., Xu, B., Warde-Farley, D., Ozair, S., ... Bengio, Y. (2014). Generative adversarial nets advances in neural information processing systems. *arXiv preprint arXiv:1406.2661*.
10. Hussein, R., Ahmed, M. O., Ward, R., Wang, Z. J., Kuhlmann, L., & Guo, Y. (2019). Human intracranial EEG quantitative analysis and automatic feature learning for epileptic seizure prediction. *arXiv preprint arXiv:1904.03603*.
11. Irvin, J., Rajpurkar, P., Ko, M., Yu, Y., Ciurea-Ilcus, S., Chute, C., ... others (2019). Chexpert: A large chest radiograph dataset with uncertainty labels and expert comparison. In *Proceedings of the AAAI Conference on Artificial Intelligence* (Vol. 33, pp. 590–597). Part of the PKP Publishing Services Network.
12. Islam, J., & Zhang, Y. (2018). Brain MRI analysis for Alzheimer's disease diagnosis using an ensemble system of deep convolutional neural networks. *Brain Informatics*, 5, 1–14.
13. Kavitha, M., Srinivas, P., Kalyampudi, P. L., Srinivasulu, S., et al. (2021). Machine learning techniques for anomaly detection in smart healthcare. In *2021 Third International Conference on Inventive Research in Computing Applications (ICIRCA)* (pp. 1350–1356). IEEE.
14. Le, M. H., Chen, J., Wang, L., Wang, Z., Liu, W., Cheng, K.-T. T., & Yang, X. (2017). Automated diagnosis of prostate cancer in multi-parametric MRI based on multimodal convolutional neural networks. *Physics in Medicine & Biology*, 62(16), 6497.
15. Lu, D., Popuri, K., Ding, G. W., Balachandar, R., & Beg, M. F. (2018). Multimodal and multiscale deep neural networks for the early diagnosis of Alzheimer's disease using structural MR and FDG-PET images. *Scientific Reports*, 8(1), 5697.
16. Melillo, P., Izzo, R., Orrico, A., Scala, P., Attanasio, M., Mirra, M., ... Pecchia, L. (2015). Automatic prediction of cardiovascular and cerebrovascular events using heart rate variability analysis. *PloS One*, 10(3), e0118504.

17. Ouyang, D., He, B., Ghorbani, A., Yuan, N., Ebinger, J., Langlotz, C. P., ... others (2020). Video-based AI for beat-to-beat assessment of cardiac function. *Nature*, 580(7802), 252–256.
18. Penzel, T., Moody, G. B., Mark, R. G., Goldberger, A. L., & Peter, J. H. (2000). The Apnea-ECG database. In *Computers in cardiology 2000. vol. 27 (cat. 00ch37163)* (pp. 255–258). IEEE.
19. Risacher, S., Kim, S., Shen, L., Nho, K., Foroud, T., Green, R., ... others (2013). The role of apolipoprotein E (APOE) genotype in early mild cognitive impairment (E-MCI). *Front Aging Neurosci*, 5, 11.
20. Šabić, E., Keeley, D., Henderson, B., & Nannemann, S. (2021). Healthcare and anomaly detection: Using machine learning to predict anomalies in heart rate data. *AI & Society*, 36(1), 149–158.
21. Shehata, M., Khalifa, F., Soliman, A., Ghazal, M., Taher, F., Abou El-Ghar, M., ... El-Baz, A. (2018). Computer-aided diagnostic system for early detection of acute renal transplant rejection using diffusion-weighted MRI. *IEEE Transactions on Biomedical Engineering*, 66(2), 539–552.
22. Tigaran, S., Mølgaard, H., McClelland, R., Dam, M., & Jaffe, A. (2003). Evidence of cardiac ischemia during seizures in drug refractory epilepsy patients. *Neurology*, 60(3), 492–495.
23. Truong, N. D., Nguyen, A. D., Kuhlmann, L., Bonyadi, M. R., Yang, J., & Kavehei, O. (2017). A generalised seizure prediction with convolutional neural networks for intracranial and scalp electroencephalogram data analysis. *arXiv preprint arXiv:1707.01976*.
24. Tumuluru, P., Lakshmi, C. P., Sahaja, T., & Prazna, R. (2019). A review of machine learning techniques for breast cancer diagnosis in medical applications. In *2019 3rd International Conference on I-SMAC (IoT in Social, Mobile, Analytics and Cloud)(I-SMAC)* (pp. 618–623). IEEE.
25. Ukil, A., Bandyoapdhyay, S., Puri, C., & Pal, A. (2016). IoT healthcare analytics: The importance of anomaly detection. In *2016 IEEE 30th International Conference on Advanced Information Networking and Applications (AINA)* (pp. 994–997). IEEE.
26. Wang, X., Peng, Y., Lu, L., Lu, Z., Bagheri, M., & Summers, R. M. (2017). Chestx-ray8: Hospital-scale chest X-ray database and benchmarks on weakly-supervised classification and localization of common thorax diseases. In *Proceedings of the IEEE Conference on Computer Vision and Pattern Recognition* (pp. 2097–2106). IEEE.

27. Yang, X., Qi, X., & Zhou, X. (2023). Deep learning technologies for time series abnormality detection in healthcare: A review. *IEEE Access*, 11, 117788–117799.
28. Yoo, Y., Tang, L. Y., Brosch, T., Li, D. K., Kolind, S., Vavasour, I., … Tam, R. C. (2018). Deep learning of joint myelin and T1w MRI features in normal-appearing brain tissue to distinguish between multiple sclerosis patients and healthy controls. *NeuroImage: Clinical*, 17, 169–178.
29. Zeng, L.-L., Wang, H., Hu, P., Yang, B., Pu, W., Shen, H., … others (2018). Multi-site diagnostic classification of schizophrenia using discriminant deep learning with functional connectivity MRI. *EBioMedicine*, 30, 74–85.
30. Zhang, Z., & Sabuncu, M. (2018). Generalized cross entropy loss for training deep neural networks with noisy labels. *Advances in Neural Information Processing Systems*, 31, 1–11.
31. Zhao, B., James, L. P., Moskowitz, C. S., Guo, P., Ginsberg, M. S., Lefkowitz, R. A., … Schwartz, L. H. (2009). Evaluating variability in tumor measurements from same-day repeat CT scans of patients with non–small cell lung cancer. *Radiology*, 252(1), 263–272.

CHAPTER 4

Safeguarding Health in the Cloud

A Cutting-Edge Machine Learning Intrusion Detection

P.K. Monika, Bi Bi Hajira, and Pallavi Joshi

4.1 INTRODUCTION

The growing digitisation of healthcare institutions, combined with the widespread adoption of cloud technology, has certainly altered the data management environment, with more individuals utilising cloud technology and encountering online risks like false emails and malware, as occurred particularly during COVID-19. Normal antivirus software is insufficient to protect against smart assaults (Mukhopadhyay, Mega, and DR, 2021). Mukhopadhyay and Prajwal (2021) introduce a new approach for detecting hidden infections and bogus websites, making them more secure online. While enhanced accessibility, collaboration, and cost-effectiveness are obvious benefits, there is a looming concern regarding the security of critical healthcare data. It has never been more crucial to protect the confidentiality and availability of patient data stored in the cloud (Mukhopadhyay and Srinivas, 2023). Healthcare organisations' exposure to cyber threats grows as they manage this paradigm transition, necessitating novel security solutions. A paper by Vandana et al. (2020) conveys a hybrid cryptographic strategy for safe medical information transfer that incorporates both symmetric and asymmetric algorithms, despite possible constraints in the scalability and efficiency of cloud-based services such as Amazon EC2 and S3 for handling sensitive medical data. In response

DOI: 10.1201/9781003463559-4

to this necessity, a study by Joshi and Raghuvanshi (2022) begins a journey to methodically develop and implement an intrusion detection system (IDS) optimised for safe cloud applications in healthcare, with a special emphasis on leveraging advanced machine learning (ML) approaches. The GMKFDA-MI model, which does not include anomaly detection or optimisation techniques, is recommended by the authors as a solution to the problems associated with Internet of Things (IoT) networks, which include real-time data processing, accurate event prediction, energy-efficient aggregation, scalability, and flexibility. The paper by Srinivas, Mohan, and Joshi (2023) examines the evolution of wireless sensor networks (WSNs) and the growing IoT from 2005 to 2020, addressing difficulties such as anomaly detection, optimisation tactics, and machine learning integration. For WSN and IoT applications, the authors promote lightweight models for increased intelligence and dependability. To protect the world of the Internet of Medical Things (IoMT), a study by Si-Ahmed, Al-Garadi, and Boustia (2023) focuses on the installation of machine learning-based IDS, highlighting the benefits, limitations, and challenges. To solve these problems, the authors admit that further research is required. The paper by Attou, Guezzaz, Benkirane, Azrour, and Farhaoui (2023) describes a cloud-based system for intrusion detection that achieves high accuracy on the Bot-IoT and NSL-KDD datasets by combining feature optimisation with random forest (RF). However, recall performance needs to be increased using additional ensemble approaches.

Numerous studies attest to the importance of implementing intrusion detection mechanisms as a critical layer of defence against malicious activities. However, as healthcare systems gravitate towards cloud infrastructures, the literature exposes a gap in the depth of exploration into the specific challenges associated with securing healthcare data in the cloud. While existing research acknowledges the need for adaptive and sophisticated solutions, it often lacks granularity in addressing the unique nuances of diverse healthcare data types, regulatory compliance, and the critical need for real-time threat detection in the cloud.

The primary contributions of this chapter are as follows:

(1) In our study, we used Isolation Forest for feature selection in the healthcare cloud; it effectively finds anomalies by prioritising outlier identification in different behaviours.

(2) Our method combines random forest and decision tree classifiers after Isolation Forest anomaly detection to effectively handle a wide range of healthcare cloud security threats.

(3) By reducing dimensionality, principal component analysis (PCA) increases proactive threat detection in healthcare cloud security by facilitating data visualisation for stakeholders to comprehend assault distributions and patterns.

The remaining sections of our work will be structured as follows: Section 4.2 examines literature on healthcare in the cloud. Section 4.3 describes the methodology and recommended techniques to counter the issues addressed in the problem statement. It also depicts the overall flow schema of the model. Section 4.4 summarises the simulation study of the proposed model using the two datasets. Finally, the article concludes in Section 4.5.

4.2 RELATED WORKS

A paper by Mukhopadhyay, Remanidevi Devidas, Rangan, and Ramesh (2024) discusses rural healthcare difficulties and provides IoT-based solutions for real-time patient monitoring. However, it does not investigate alternate communication technologies and instead focuses primarily on simulation-based experimentation. A paper by Kumar et al. (2022) proposes FMMNN-IDS, a deep learning-based intrusion detection system, addressing threats in IoT, the cloud, e-commerce, banking, and healthcare. It outperforms existing approaches, emphasising improved intrusion detection accuracy. A study by Sundas, Badotra, Bharany, Almogren, Tag-ElDin, and Rehman (2022) describes HealthGuard, a machine learning-based security architecture that exhibits excellent efficacy and accuracy in monitoring and identifying possible cyber threats in smart healthcare systems. A paper by Patil et al. (2022) highlights the value of interpretability and transparency by presenting an intrusion detection system that makes use of machine learning ensemble techniques, including support vector machine (SVM), random forest, and decision trees. A paper by Iwendi et al. (2021) improve smart healthcare security through a machine learning system, combining random forest and genetic algorithms for intrusion detection, emphasising feature optimisation in Security of Things (SoT). A paper by Sadaf and Sultana (2020) addresses fog computing's security challenges, proposing Auto-IF for intrusion detection using deep learning. It reviews IoT security, fog computing studies, and intrusion detection methods in fog environments. A paper by Singh, Gaba, Kaur, Hedabou, and Gurtov (2022) addresses IoT-enabled medical device security, proposing a Dew-Cloud model with hierarchical federated learning. It highlights IoMT vulnerabilities, emphasising the need for enhanced security measures. The increasing attack surface in smart healthcare systems is addressed by researchers who suggest a decentralised, predictive method for detecting and blocking malicious traffic in IoMT networks utilising Bluetooth, AI, and deep learning (Zubair et al., 2022). A paper by Narayan, Mookherji, Odelu, Prasath, Turlapaty, and Das (2023) emphasises IoT security in healthcare, proposing a random forest-based model using the CICIoT2023 dataset. It enhances precision, recall, and F1 score, discussing imbalanced datasets and F1's critical role in sensitive applications. A paper by Lipsa and Dash (2023) focuses on securing digital twins (DT) using an IDS. It introduces a hybrid model integrating deep learning and random forest, demonstrating higher accuracy and lower computing time than other models. A paper by Hernandez-Jaimes, Martinez-Cruz, Ramírez-Gutiérrez, and Feregrino-Uribe (2023) investigates security challenges in IoMT, introducing a novel intrusion detection taxonomy with comparative methods analysis, dataset classification, and discussions on cybersecurity, AI, legal aspects, and IoMT challenges. A paper by Badri (2023) addresses healthcare data security challenges in the IoT era, proposing an MCERF-HPS intrusion detection system with blockchain integration. It emphasises the significance of intrusion detection in healthcare applications. A paper by Tauqeer, Iqbal, Ali, Zaman, and Chaudhry (2022) address IoT security, specifically in the healthcare sector (IoMT), using ML algorithms (RF, Gradient Boosting [GB], SVM) to detect cyberattacks. Evaluation of the WUSTL EHMS 2020 dataset shows superior performance. A paper by Subbiah, Anbananthen, Thangaraj, Kannan, and Chelliah (2022) addresses wireless sensor network (WSN) security using machine learning-based intrusion detection. It proposes a novel (BFS-GSRF) algorithm, outperforming existing methods. A paper by Subasi, Algebsani, Alghamdi, Kremic, Almaasrani, and Abdulaziz (2021) addresses cybersecurity challenges in smart healthcare within a smart city, emphasising IoT-based systems. It proposes a bagging ensemble classifier, achieving superior intrusion detection performance. A paper by Zhang, Zulkernine, and Haque (2008) introduces systematic frameworks employing the random forest algorithm for misuse, anomaly, and hybrid intrusion detection systems. It addresses challenges in feature selection and imbalanced intrusion, and proposes effective detection approaches. A paper by Hady, Ghubaish, Salman, Unal, and Jain (2020) presents an IoT-based Enhanced Healthcare Monitoring System (EHMS) using machine learning for intrusion detection. Combining network flow and biometric data improves detection performance by 7%–25%. A paper by Gupta and Jiwani (n.d.) addresses growing concerns about healthcare cybersecurity and proposes the use of a bagging ensemble classifier, with a focus on improving infrastructure for healthcare security. A review paper by Azam, Islam, and Huda (2023) examines recent advancements in IDS taxonomy, techniques, and evaluation datasets. It emphasises the challenges posed by attackers, explores machine learning and deep learning-based IDS, and proposes a decision tree model for effective detection. Based on theoretical efficiency, a study by Xiang et al. (2023) presents an optimisation approach for isolation tree structure called OptIForest, which outperforms state-of-the-art methods such as deep learning in anomaly detection. A paper by Phang (2022) proposes using a generative adversarial network to create an active system to detect intrusions for IoT security, with a focus on medical equipment.

4.3 METHODOLOGY

4.3.1 Problem Statement

Due to the insecurity of cloud infrastructures against future cyber threats, the proliferation of cloud services in healthcare presents a security challenge. To address this, we proposed an IDS for cloud-based healthcare systems that utilises machine learning and provides real-time threat detection. The objective of this solution is to improve security by ensuring the confidentiality and integrity of data in relation to patients as well as compliance with legislation on privacy.

4.3.2 Proposed Methodology

4.3.2.1 Isolation Forest

Isolation Forest is a method used to detect outliers or abnormalities in datasets. Isolation Forest rapidly finds anomalies by subsampling data and calculating average path lengths, making it appropriate for high-dimensional datasets without the need for specialised data distributions.

It excels in identity theft detection and network security because of its noise resistance, low parameter tuning requirements, and adaptability in circumstances with little labelled data.

4.3.2.2 Random Forest Algorithm

Random forest is a widely recognised ensemble learning approach for both regression and classification applications that produces several decision trees from training data and random attributes. It is useful for managing high-dimensional data because it encourages variety, reduces overfitting, and increases model resilience. Due to its ease of use, scalability, and superior performance in both classification and regression tasks, it is extensively employed in a broad range of applications, including image identification, healthcare, and finance. It also enhances model resilience, which makes it useful for managing high-dimensional data.

4.3.2.3 Decision Tree

A decision tree is an algorithmic learning approach for classification and regression issues. Based on the characteristics that best distinguish the target variable, it divides the dataset into subgroups. Every division is represented by a node in the tree, with branches denoting potential feature values, and leaf nodes holding the anticipated results. Decision-making trees handle both category and numerical data with ease, are robust against outliers, and are straightforward to use.

4.3.2.4 Principal Component Analysis (PCA)

Principal component analysis is a significant dimensionality reduction approach used in data analysis and machine learning that converts original variables into orthogonal components to identify underlying patterns. These components, sorted by variance collected, help to analyse and visualise high-dimensional datasets while minimising information loss. PCA reduces multicollinearity and has been widely used in genomics, image processing, signal analysis, and finance to discover complex patterns and correlations, allowing for more efficient data exploration and interpretation in a variety of sectors.

4.3.2.5 Support Vector Machine (SVM)

SVM is a powerful supervised algorithm for learning which allows for both classification and regression. It performs by determining the best hyperplane for separating data points into discrete classes while increasing the gap between them. SVM can handle both linear and nonlinear data, works well in high-dimensional settings, and is resistant to overfitting. SVM requires precise hyperparameter selection and can be computationally expensive for large datasets. Regardless, SVM is employed in a variety of applications.

4.3.2.6 k-Nearest Neighbours (k-NN)

k-nearest neighbours is a method of machine learning used in problems involving regression as well as classification. It generates class labels by computing distances to *k*-nearest neighbours using metrics like Manhattan or Euclidean distance. *k*-NN is non-parametric in nature and adapts well to a variety of data forms, but at the price of processing resources, particularly for big datasets. Choosing a suitable *k* number is crucial since it affects both algorithm performance and applicability for various applications.

4.3.3 Proposed System Architecture

In Figure 4.1, the method uses machine learning algorithms, preprocessing, and feature engineering to find anomalies. The anomalies are then assessed using performance measures.

(1) **Preprocessing**: To optimise the performance of the model, the methods of feature engineering, encoding, and cleaning are applied to the NSL-KDD and UNSW-NB15 datasets.

(2) **Data splitting**: The splitting of preprocessed datasets into training (80%) and testing (20%) subsets ensures that the model is assessed on unobserved data.

(3) **Model training**: Using the training set of data, models are trained for anomaly detection using a number of machine learning approaches that involve ensemble methods.

(4) **Testing phase**: Using the testing data, trained models are employed to forecast anomalies and distinguish between anomalous and typical network connections.

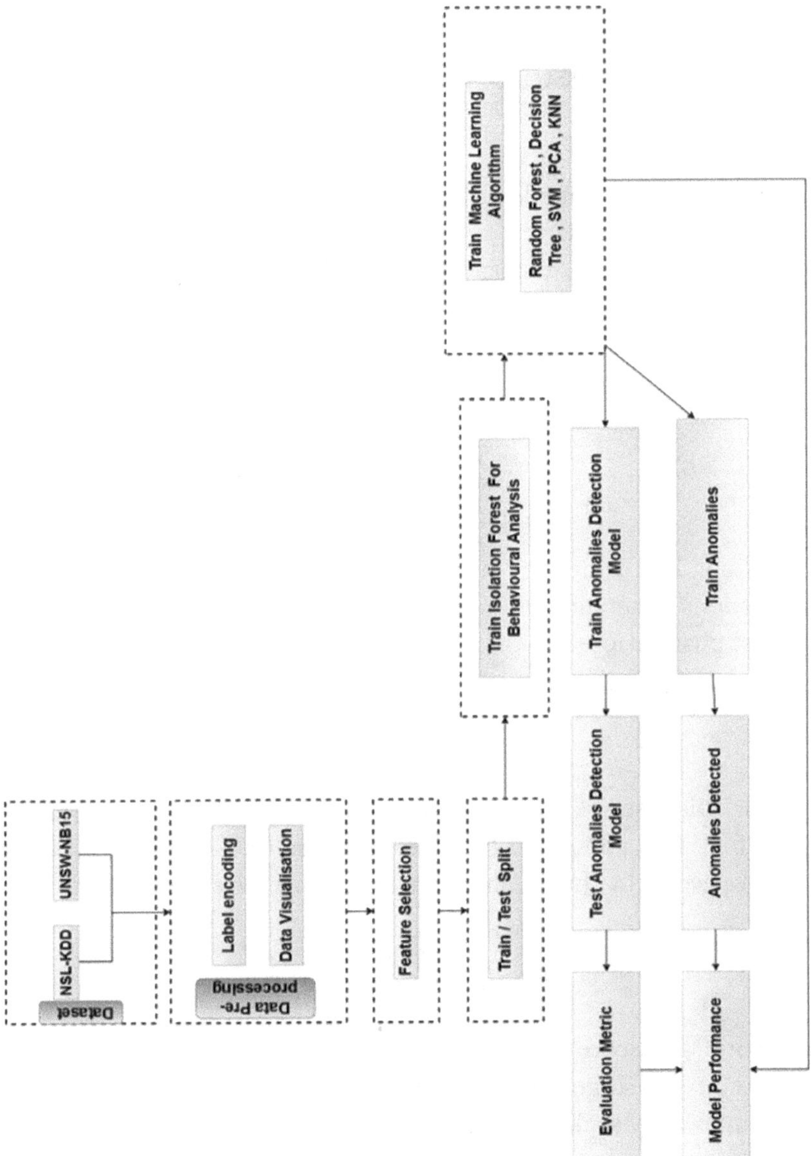

FIGURE 4.1 Proposed system architecture.

(5) **Performance evaluation**: Evaluation metrics analyse the performance of a model and give information into its capacity to detect anomalous network behaviour.

4.4 RESULTS AND DISCUSSION

4.4.1 Evaluation Metrics

The effectiveness of machine learning models is determined by evaluation measures. They help evaluate a model's efficiency in executing a certain job, such as regression or classification. Some of the common measures for evaluation are outlined as follows:

(1) Accuracy: This is calculated as the proportion of accurately predicted outcomes to all of the predictions the model generates.

$$Accuracy = \frac{\text{Total Accurate Predictions}}{\text{Total Number of Predictions}} \tag{4.1}$$

(2) Precision: This is derived by reducing the number of true positive predictions generated by the predictive algorithm by the number of correct positive forecasts.

$$Precision = \frac{TP + FP}{TP} \tag{4.2}$$

(3) Recall: This assesses the accuracy with which authentic positives are identified. It is computed by combining the total number of actual positive events by the number of actually positive estimations.

$$Recall = \frac{TP + FN}{TP} \tag{4.3}$$

(4) F1 score: The score for F1 is a ratio between harmonics for accuracy and recall. It generates one data point that sums up the model's performance by striking a balance between recall and accuracy.

$$F1 - Score = \frac{2 \times Precision \times Recall}{Precision + Recall} \tag{4.4}$$

TABLE 4.1 Summary of datasets

Dataset	No. of attributes	Instances	No. of classes
NSL-KDD	41	125973	2
UNSW-NB15	49	2540044	2

TABLE 4.2 Performance metrics for proposed methods on NSL-KDD data

Method	Accuracy	Precision	Recall	F1 score
Random forest	98	97	76	85
Decision tree	99	95	83	89
PCA	94	83	14	24
SVM	90	90	90	90
k-NN	80	80	81	81

TABLE 4.3 Performance metrics of proposed methods on UNSW-NB15 data

Proposed method	Accuracy	Precision	Recall	F1 score
Random forest	94.95	95.95	94.95	95.93
Decision tree	94.89	94.89	94.90	94.90
PCA	94.28	94.28	94.26	94.27
SVM	93.52	93.52	93.90	93.30
k-NN	94.60	94.60	95.48	94.53

(5) Confusion matrix: A visual representation known as a confusion matrix is used to describe how well a classification system performs. It provides a detailed analysis of the predictions made by the model compared to what actually happens in a range of domains. This matrix helps in evaluating the model's accuracy, identifying patterns in misclassification, and understanding its overall importance in classification tasks.

4.4.2 Discussion

In this experiment, we are using two datasets: the UNSW-NB15 and the NSL-KDD. Details of these are presented in Table 4.1.

Table 4.2 and Table 4.3 display the metrics for accuracy, F1 score, recall, and precision, providing an overview of the results that have been achieved for the UNSW-NB15 and NSL-KDD datasets.

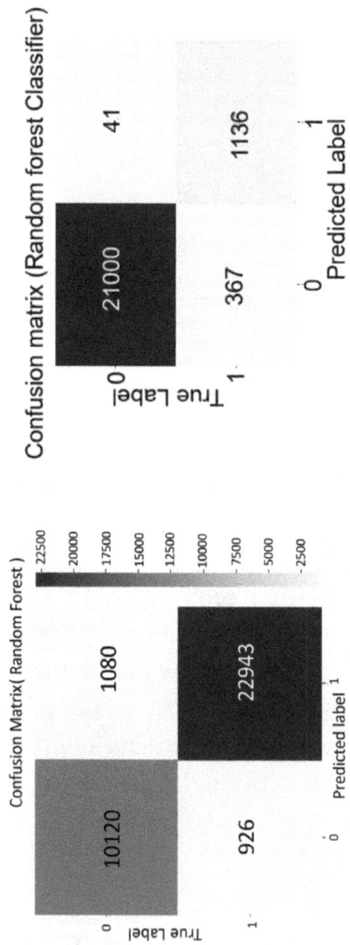

FIGURE 4.2 Comparison of random forest models on different datasets.

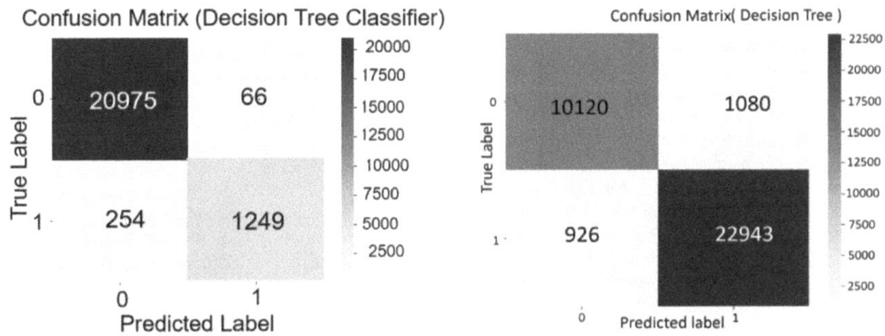

FIGURE 4.3 Comparison of decision tree models on different datasets.

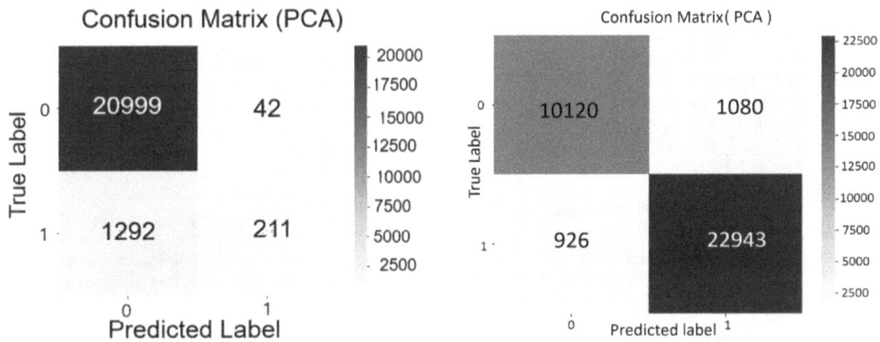

FIGURE 4.4 Comparison of PCA on different datasets.

4.4.3 Confusion Matrix for Both Datasets

After combining the confusion matrices from both datasets, it gives high precision (95.95%) with low false positives and negatives (4.05%), which is shown in Figure 4.2 (a) and (b).

From Figure 4.3 (a) and (b), the metric shows that the decision tree classifier achieved 98.58% accuracy and is good at detecting complicated data. The confusion matrix (94.89%) is quite accurate.

PCA has reduced accuracy (94.08%) because of its decreased dimensionality. The confusion matrix demonstrates great accuracy (94.28%) with balanced true positives and negatives, as shown in Figure 4.4 (a) and (b).

In Figure 4.5 (a) and (b), the test accuracy is 90%, representing balanced class metrics. The confusion matrix accuracy is 93.52%, with balanced true positives and negatives.

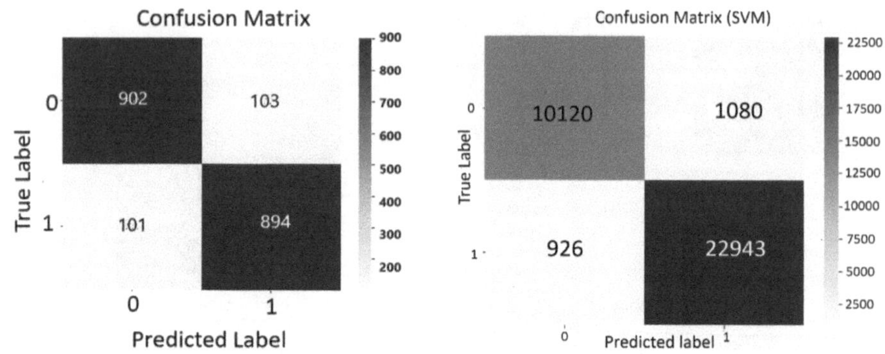

FIGURE 4.5 Comparison of SVM on different datasets.

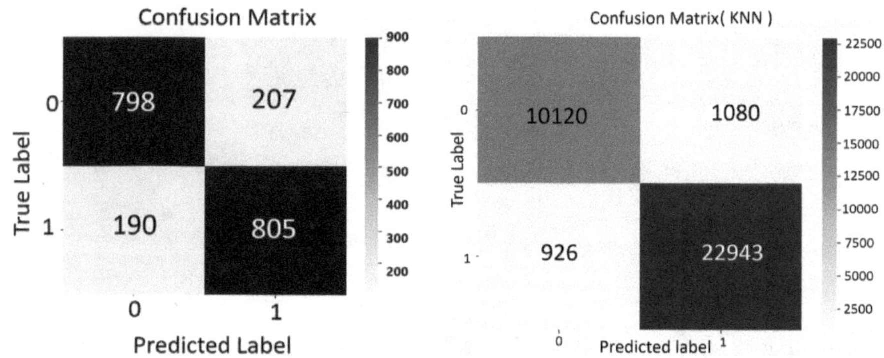

FIGURE 4.6 Comparison of *k*-NN on different datasets.

In Figure 4.6 (a) and (b), the *k*-NN achieved 80% test accuracy with balanced evaluation metrics for both classes. The matrix shows high accuracy (94.60%) with balanced true positive and true negative rates.

4.5 CONCLUSION AND FUTURE SCOPE

Our research provides a comprehensive method for increasing the security of cloud-based healthcare systems by developing an upgraded IDS that employs a variety of machine learning approaches. We obtained positive results in real-time threat detection and anomaly classification by merging machine learning models, as seen by our examination of both datasets. Prospective studies will enhance healthcare system security through dynamic model adaptation, deep learning, adversarial attack resilience, and interoperability.

REFERENCES

1. Attou, H., Guezzaz, A., Benkirane, S., Azrour, M., & Farhaoui, Y. (2023). Cloud-based intrusion detection approach using machine learning techniques. *Big Data Mining and Analytics*, 6(3), 311–320.
2. Azam, Z., Islam, M. M., & Huda, M. N. (2023). Comparative analysis of intrusion detection systems and machine learning based model analysis through decision tree. *IEEE Access*, 11, 80348–80391.
3. Badri, S. (2023). HO-CER: Hybrid-optimization-based convolutional ensemble random forest for data security in healthcare applications using Blockchain technology. *Electronic Research Archive*, 31(9), 5466–5484.
4. Gupta, K., & Jiwani, N. (n.d). Cybersecurity framework in healthcare sector and techniques to mitigate and detect attacks. *Journal of Xidian University*, 16(9), 516–521.
5. Hady, A. A., Ghubaish, A., Salman, T., Unal, D., & Jain, R. (2020). Intrusion detection system for healthcare systems using medical and network data: A comparison study. *IEEE Access*, 8, 106576–106584. Hernandez-Jaimes, M. L., Martinez-Cruz, A., Ramírez-Gutiérrez, K. A., & Feregrino-Uribe, C. (2023). Artificial Intelligence for IOMT security: A review of intrusion detection systems, attacks, datasets and cloud-fog-edge architectures. *Internet of Things*, 23, October 2023, 100887.
6. Iwendi, C., et al. (2021). Security of things intrusion detection system for smart healthcare. *Electronics*, 10(12), 1375.
7. Joshi, P., & Raghuvanshi, A. S. (2022). A dual synchronization prediction-based data aggregation model for an event monitoring IoT network. *Journal of Intelligent & Fuzzy Systems*, 42(4), 3445–3464.
8. Kumar, A., Umurzoqovich, R. S., Duong, N. D., Kanani, P., Kuppusamy, A., Praneesh, M., & Hieu, M. N. (2022). An intrusion identification and prevention for cloud computing: From the perspective of deep learning. *Optik*, 270, 170044.

9. Lipsa, S., & Dash, R. K. (2023). A novel intrusion detection system based on deep learning and random forest for digital twin on IoT platform. *International Journal of Engineering Research & Technology*, 2, 51–64.
10. Mukhopadhyay, A., Mega, S., & DR, B. K. (2021). SGBEHIM: Smart GIS based emergency healthcare infrastructure mapping. In *2021 3rd International Conference on Inventive Research in Computing Applications (ICIRCA)* (pp. 981–986). IEEE.
11. Mukhopadhyay, A., & Prajwal, A. (2021). EDITH—a robust framework for prevention of cyber attacks in the Covid era. In *2021 2nd International Conference for Emerging Technology (INCET)* (pp. 1–8). IEEE.
12. Mukhopadhyay, A., Remanidevi Devidas, A., Rangan, V. P., & Ramesh, M. V. (2024). A QoS-aware IoT edge network for mobile telemedicine enabling in-transit monitoring of emergency patients. *Future Internet*, 16(2), 52.
13. Mukhopadhyay, A., & Srinivas, S. (2023). Regression for predicting COVID-19 infection possibility based on underlying cardiovascular disease: A medical score-based approach. In *International Conference on Communications and Cyber Physical Engineering 2018* (pp. 679–691). Springer.
14. Narayan, K., Mookherji, S., Odelu, V., Prasath, R., Turlapaty, A. C., & Das, A. K. (2023). IIDS: Design of intelligent intrusion detection system for internet-of-things applications. *arXiv preprint arXiv:2308.00943*.
15. Patil, S., Varadarajan, V., Mazhar, S. M., Sahibzada, A., Ahmed, N., Sinha, O., … Kotecha, K. (2022). Explainable Artificial Intelligence for intrusion detection system. *Electronics*, 11(19), 3079.
16. Phang, J. J. H. (2022). *Real-time intrusion detection system in IoT medical devices* [Unpublished doctoral dissertation]. UTAR.
17. Sadaf, K., & Sultana, J. (2020). Intrusion detection based on autoencoder and isolation forest in fog computing. *IEEE Access*, 8, 167059–167068. Si-Ahmed, A., Al-Garadi, M. A., & Boustia, N. (2023). Survey of machine learning based intrusion detection methods for internet of medical things. *Applied Soft Computing*, 140, 110227. https://www.sciencedirect.com/science/article/pii/S1568494623002454

18. Singh, P., Gaba, G. S., Kaur, A., Hedabou, M., & Gurtov, A. (2022). Dew-cloud-based hierarchical federated learning for intrusion detection in IoMT. *IEEE Journal of Biomedical and Health Informatics*, 27(2), 722–731.
19. Srinivas, H. R., Mohan, P., & Joshi, P. (2023). Detection of energy efficient sensor node in EHWSN. In *2023 World Conference on Communication & Computing (WCONF)* (pp. 1–6). IEEE.
20. Subasi, A., Algebsani, S., Alghamdi, W., Kremic, E., Almaasrani, J., & Abdulaziz, N. (2021). Intrusion detection in smart healthcare using bagging ensemble classifier. In *International Conference on Medical and Biological Engineering* (pp. 164–171). IEEE.
21. Subbiah, S., Anbananthen, K. S. M., Thangaraj, S., Kannan, S., & Chelliah, D. (2022). Intrusion detection technique in wireless sensor network using grid search random forest with Boruta feature selection algorithm. *Journal of Communications and Networks*, 24(2), 264–273.
22. Sundas, A., Badotra, S., Bharany, S., Almogren, A., Tag-ElDin, E. M., & Rehman, A. U. (2022). Healthguard: An intelligent healthcare system security framework based on machine learning. *Sustainability*, 14(19), 11934.
23. Tauqeer, H., Iqbal, M. M., Ali, A., Zaman, S., & Chaudhry, M. U. (2022). Cyberattacks detection in IoMT using machine learning techniques. *Journal of Computing & Biomedical Informatics*, 4(01), 13–20.
24. Vandana, R., BJ, S. K., et al. (2020). Integrity based authentication and secure information transfer over cloud for hospital management system. In *2020 4th International Conference on Intelligent Computing and Control Systems (ICICCS)* (pp. 139–144). IEEE.
25. Xiang, H., Zhang, X., Hu, H., Qi, L., Dou, W., Dras, M., … Xu, X. (2023). Optiforest: Optimal isolation forest for anomaly detection. *arXiv preprint arXiv:2306.12703*.
26. Zhang, J., Zulkernine, M., & Haque, A. (2008). Random-forests-based network intrusion detection systems. *IEEE Transactions on Systems, Man, and Cybernetics, Part C (Applications and Reviews)*, 38(5), 649–659.
27. Zubair, M., Ghubaish, A., Unal, D., Al-Ali, A., Reimann, T., Alinier, G., … Qadir, J. (2022). Secure Bluetooth communication in smart healthcare systems: A novel community dataset and intrusion detection system. *Sensors*, 22(21), 8280.

CHAPTER 5

Deep Learning for Anomaly Detection in IoT Time Series

Jiange Jiang and Chen Chen

5.1 INTRODUCTION

In the Internet of Things (IoT), interconnected devices generate vast amounts of time-series data. This data provides insights into operational performance, environmental conditions, and human behaviour. Effective time-series anomaly detection is crucial for identifying deviations from expected patterns within these sequences of data points. Across domains such as finance (Moschini, Houssou, Bovay, & Robert-Nicoud, 2021; Hemdan & Manjaiah, 2022), industrial manufacturing (Y. Liu et al., 2020; Kong, Li, Jiang, Wang, & Song, 2021), healthcare (Wolleb, Bieder, Sandkühler, & Cattin, 2022), and cybersecurity (Rashid, Ahmed, Sikos, & Haskell-Dowland, 2022; Garg, Zhang, Samaran, Savitha, & Foo, 2021), where time-series data pervades every facet of operations, understanding and detecting anomalies within these data streams is crucial for ensuring the integrity, reliability, and security of systems and processes. For instance, in sectors like manufacturing and healthcare, where system reliability is paramount, real-time anomaly detection ensures the smooth operation of processes (Hao, Yang, & Yang, 2021), reduces downtime, and enhances overall productivity. Furthermore, by promptly detecting anomalies, businesses can mitigate potential risks such as fraud, equipment failures, network intrusions (Mothukuri, Khare, Parizi, Pouriyeh, Dehghantanha, & Srivastava, 2021), or health complications, thereby minimising financial losses and

DOI: 10.1201/9781003463559-5

operational disruptions. However, the inherent temporal correlation in time-series renders anomaly detection more challenging compared to image anomaly detection. Firstly, time-series data often contains high levels of noise and volatility, including various forms of random disturbances or periodic fluctuations, making it more difficult to identify anomalies accurately within the data. Furthermore, anomalies in time-series data are typically rare events relative to normal conditions, leading to data imbalance, which makes models more susceptible to the influence of normal data, thereby reducing the accuracy and reliability of anomaly detection. Lastly, some anomalies may arise from long-term dependencies, necessitating models that can effectively capture and understand long-term sequence patterns. However, time-series data often suffers from missing or incomplete data, which may result from sensor failures, communication interruptions, or data recording errors. Addressing this data incompleteness requires appropriate techniques for data imputation or interpolation to ensure the accuracy and reliability of anomaly detection. The traditional methods for anomaly detection are usually based on statistical principles and machine learning algorithms, which learn patterns and rules from time-series data (J. Li, Izakian, Pedrycz, & Jamal, 2021). However, there are several limitations when applied to anomaly detection in IoT time series. Firstly, these methods often require manual feature engineering (J. E. Zhang, Wu, & Boulet, 2021), where domain expertise is necessary to select and engineer relevant features from the raw data. This process can be time-consuming, labour-intensive, and may not fully capture the complex temporal patterns present in IoT time-series data. Secondly, traditional machine learning algorithms typically assume that data instances are independent and identically distributed (i.i.d.), which may not hold true for time-series data where sequential dependencies exist (Mestav, Wang, & Tong, 2022; Pang, Cao, & Aggarwal, 2021). This can lead to suboptimal performance and difficulty in capturing the temporal dynamics inherent in IoT environments. Additionally, traditional machine learning methods may struggle with handling high-dimensional and heterogeneous time-series data, such as those generated by diverse sensors in IoT systems (Apostol, Truică, Pop, & Esposito, 2021). These methods may not effectively capture the diverse characteristics and correlations present in such data, leading to reduced accuracy in anomaly detection. The emergence of deep learning techniques has revolutionised time-series anomaly detection, addressing many of these challenges. Deep learning models, such as recurrent neural networks (RNNs), long short-term memory networks (LSTMs), and convolutional neural networks (CNNs), can automatically learn intricate patterns and dependencies from raw time-series data without the need for manual feature engineering. Zhang et al. (Y. Zhang, Chen, Wang, & Pan, 2021) proposed a novel deep learning-based anomaly detection algorithm, the Deep Convolutional Autoencoding Memory (CAE-M) network, which combines a deep convolutional autoencoder with a memory network to capture spatial and temporal dependencies in multi-sensor time-series data. It can achieve better performance when intra-class data exhibits similar distributions or regular patterns. Similarly, authors introduced an architecture that integrates a CNN with an LSTM-based autoencoder to enhance anomaly detection in time-series data (C. Yin, Zhang, Wang, & Xiong, 2020). These models excel at capturing complex temporal dynamics and dependencies, thus overcoming the limitations of traditional machine learning methods in IoT time-series anomaly detection. Additionally, deep learning models can handle high-dimensional and heterogeneous time-series data more effectively, leading to improved accuracy in anomaly detection tasks. As a result, the application of deep learning in time-series anomaly detection represents a promising direction for enhancing anomaly detection capabilities in IoT environments. Several survey papers have delved into the realm of anomaly detection in time series using artificial intelligence (AI) methodologies, including notable works such as the study by Cook et al. (Cook, Mısırlı, & Fan, 2019) and Li et al. (G. Li & Jung, 2023). However, existing literature often falls short in providing a holistic exploration of the nuanced challenges and advancements specific to this domain. This chapter endeavours to address this gap by conducting a meticulous review and synthesis of deep learning techniques tailored for anomaly detection in IoT time-series data. Through this endeavour, it aims to elucidate the unique intricacies and emerging trends in this interdisciplinary field, offering valuable insights to researchers, practitioners, and stakeholders alike. The remainder of this chapter is organised as follows. Section 5.2 explores the categories of time-series anomalies and their definitions, providing a comprehensive understanding of the different types of anomalies encountered in time-series data. In Section 5.3, we summarise the methods for time-series anomaly detection, encompassing both traditional statistical approaches and modern machine learning techniques. Section 5.4 focusses on the application of time-series anomaly detection in IoT. Section 5.5 examines the challenges inherent in time-series anomaly detection and proposes future research directions. Finally, Section 5.6 concludes this chapter.

5.2 TAXONOMY OF TIME-SERIES ANOMALIES

Time-series anomalies refer to events or data points within time-series data that deviate significantly from the expected patterns. These anomalies may manifest as sudden spikes, persistent irregular patterns, or significant deviations from the data distribution. We categorise time-series anomalies into three distinct types based on their characteristics and occurrence patterns: point anomalies, collective anomalies, and contextual anomalies.

5.2.1 Point Anomalies

Within the domain of time-series anomaly detection, point anomalies serve as pivotal indicators, spotlighting individual data points that veer markedly from the anticipated trajectory of the time series. These anomalies stand out for their solitary occurrences, where a sole data point demonstrates an aberration from the customary behaviour observed in the time-series data. This inherent characteristic of singularity underscores the distinctive nature of point anomalies, wherein a solitary data point exhibits a deviation from the typical pattern observed in the time-series data. Figure 5.1 shows an example of point anomalies.

From a mathematical standpoint, the identification of point anomalies entails a rigorous examination of each data point x_i within the time series $X = \{x_1, x_2, ..., x_n\}$ against a predefined threshold ϵ, against which its deviation from the expected pattern is measured. Specifically, if the absolute difference between x_i and the mean or median μ of the time series surpasses ϵ, as expressed by $|x_i - \mu| > \epsilon$, x_i, this earns the classification of a point anomaly. The significance of point anomaly detection transcends disciplinary boundaries, finding applications across diverse domains. For example, in cybersecurity, it might signify unauthorised access attempts or suspicious network activity (Huong, Bac, Long, Luong, Dan, & Thang, 2021). In financial markets, sudden fluctuations in stock prices or trading volumes could indicate anomalies warranting further investigation (Z. Xiao & Jiao, 2021).

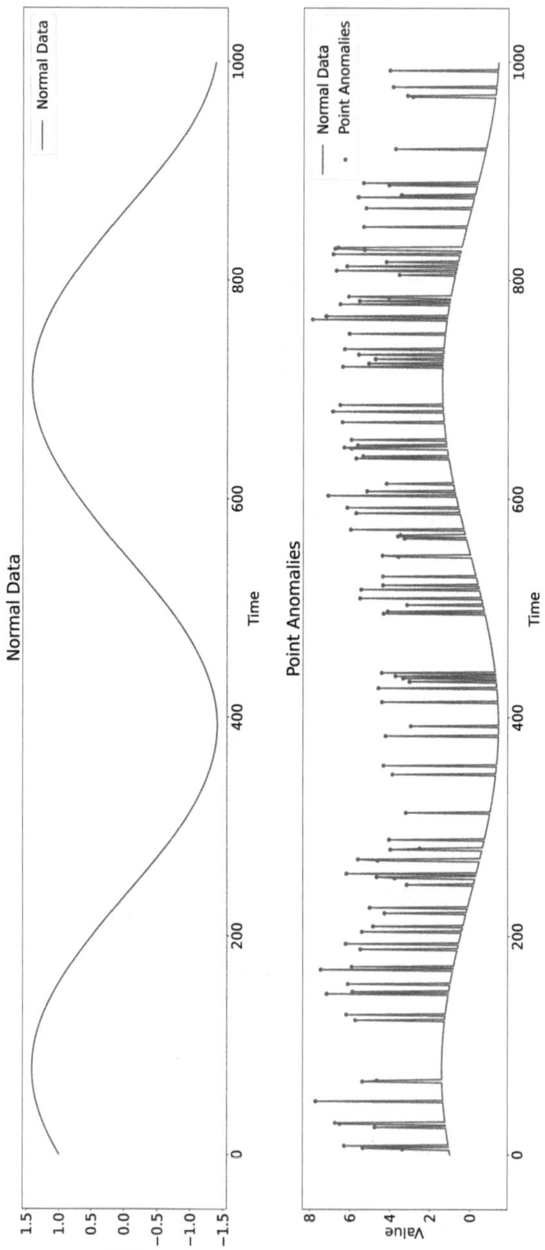

FIGURE 5.1 Example of point anomalies.

5.2.2 Collective Anomalies

Within the realm of time-series anomaly detection, collective anomalies serve as a critical focal point, shedding light on deviations from anticipated patterns within specific temporal windows or segments of the data. Unlike point anomalies, which pinpoint individual aberrant data points, collective anomalies encapsulate anomalies that manifest within defined time intervals, offering insights into broader trends or patterns of irregularity within the time series. Figure 5.2 shows an example of collective anomalies.

From a mathematical perspective, the identification of collective anomalies hinges on discerning deviations in the aggregate behaviour or statistical properties of the time-series data within a specified time interval. This involves computing the aggregated value, such as the average or sum, of data points within the interval and comparing it to the expected value. If the disparity between the aggregated value A and the expected value μ exceeds a predefined threshold ϵ, as expressed by $|A - \mu| > \epsilon$, the interval is classified as anomalous. Applications of collective anomaly detection span diverse domains, each presenting unique challenges and opportunities for analysis. For instance, in healthcare monitoring systems, anomalies in patient vital signs may alert healthcare providers to potential medical emergencies or adverse reactions to treatment (Vos, Peng, Jenkins, Shahriar, Borghesani, & Wang, 2022). Varone et al. investigated the power spectrum density (PSD), in resting-state electroencephalography (EEG), to evaluate the abnormalities in psychogenic non-epileptic seizure (PNES)-affected brains (Varone et al., 2021). The detection and analysis of collective anomalies offer valuable insights into temporal patterns of deviation within time-series data, facilitating proactive decision-making and risk management across various applications and industries.

5.2.3 Contextual Anomalies

Additionally, contextual anomalies emerge as a nuanced category, delineating instances where a data point deviates from the anticipated pattern within a specific context or subset of the time-series data. Unlike point and collective anomalies, which focus on individual data points or predefined temporal windows, contextual anomalies delve deeper into the underlying context of the data, discerning deviations within specific subsets based on contextual information. Figure 5.3 shows an example of contextual anomalies.

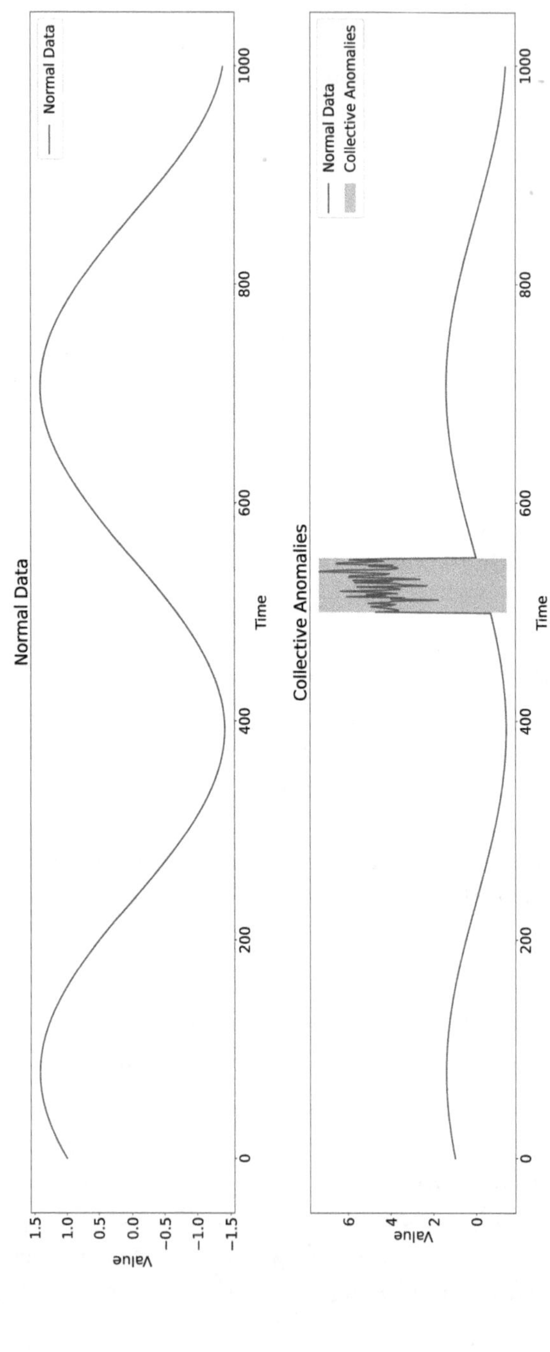

FIGURE 5.2 Example of collective anomalies.

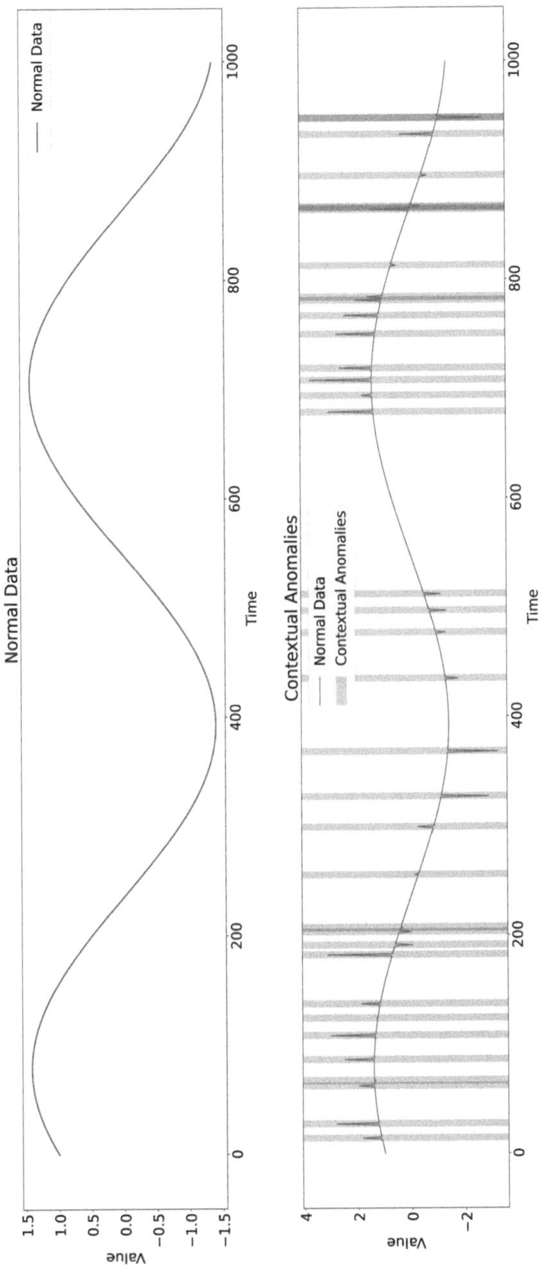

FIGURE 5.3 Example of contextual anomalies.

From a mathematical perspective, the identification of contextual anomalies hinges on scrutinising each data point x_i within the time series X against a predefined threshold ϵ, against which its deviation from the expected behaviour within a specific context or subset C is measured. Specifically, if the absolute difference between x_i and the mean or median μ_C of the subset C surpasses ϵ, as expressed by $|x_i - \mu_C| > \epsilon, x_i$, this garners the classification of a contextual anomaly. The application landscape of contextual anomaly detection spans diverse domains, where contextual information plays a pivotal role in anomaly identification and interpretation. For instance, in network security, contextual anomalies may manifest as unusual patterns of network activity within specific user groups, departments, or geographical locations, indicating potential security threats or policy violations (Pérez, Moral-Rubio, & Criado, 2021). Similarly, in supply chain management, contextual anomalies could signify irregularities in product demand or inventory levels within specific regions or market segments, necessitating adjustments in production and distribution strategies (Beteto et al., 2022). In summary, different time-series anomaly scenarios can exhibit diverse characteristics and nuances. For instance, in cybersecurity, point anomalies might indicate isolated instances of malicious access attempts or suspicious network activity, whereas collective anomalies could signify abnormal patterns of data transfer rates within specific time periods, potentially signaling network congestion or security breaches. Similarly, in financial markets, point anomalies may represent anomalous trading behaviours or market irregularities, while contextual anomalies might involve deviations in trading patterns across different geographical regions or user groups. In manufacturing, point anomalies could denote sudden deviations in machine sensor readings, indicating equipment malfunctions or production line interruptions. Meanwhile, collective anomalies might reveal prolonged periods of decreased productivity within specific shifts or production cycles, hinting at underlying process inefficiencies or resource constraints. In contrast, contextual anomalies may manifest as irregularities in product quality across different manufacturing plants or geographical locations, highlighting variations in production processes or environmental factors. Therefore, when describing the application scenarios of different anomaly types, it is essential to consider the specific requirements and nuances of each domain, ensuring that each anomaly type's cases adequately showcase their uniqueness and significance.

5.3 METHODS OF ANOMALY DETECTION

5.3.1 Statistical

Statistical-based time-series anomaly detection methods typically involve establishing a baseline model to describe the normal patterns of time-series data and then identifying anomalies based on the deviation or residuals between the data and the model.

5.3.1.1 Methods Based on Threshold

The threshold-based method relies on setting a predefined threshold, above or below which data points are classified as anomalous. Anomalies are identified based on whether they exceed or fall below this threshold. This approach assumes that anomalies exhibit significant deviations from normal data and can be captured using a straightforward threshold. Let TPR represent the true positive rate and FPR represent the false positive rate. The threshold is set to T, then:

$$\mathrm{TPR} = \frac{\mathrm{TP}}{\mathrm{TP} + \mathrm{FN}}, \tag{5.1}$$

$$\mathrm{FPR} = \frac{\mathrm{FP}}{\mathrm{FP} + \mathrm{TN}}, \tag{5.2}$$

where TP represents true positives (correctly detected anomalies), FP represents false positives (false alarms, normal data incorrectly classified as anomalies), TN represents true negatives (correctly detected normal data), and FN represents false negatives (missed anomalies). This method is commonly applied in scenarios where anomalies are expected to display extreme values compared to normal data. Akande et al. (David Akande, Kaur, Dadkhah, & Ghorbani, 2022) propose an anomaly detection approach using a threshold to discriminate between regular and aberrant log files. It finds utility in domains where anomalies are distinct and easily distinguishable from normal behaviour. Furthermore, the threshold-based method offers simplicity in implementation and understanding. It provides an intuitive approach by offering a clear distinction between normal and anomalous data points. Additionally, it demands minimal computational resources, enhancing computational efficiency. However, the method is sensitive to threshold selection, with its performance highly reliant on the chosen threshold, which may vary across different datasets. Moreover, it lacks flexibility, making it unsuitable for detecting anomalies with complex patterns or subtle deviations from normal behaviour. Additionally, it overlooks contextual information, potentially leading to false alarms or missed anomalies.

5.3.1.2 Methods Based on Time-Series Decomposition

Time-series decomposition involves breaking down a time series into its constituent components: trend, seasonality, and residual. Anomalies are typically identified in the residual component, as it represent the part of the data that cannot be explained by trend and seasonality. This method compares the residual with normal residual patterns to detect anomalies. Let $y(t)$ represent the original time series, $T(t)$ represent the trend component, $S(t)$ represent the seasonality component, and $R(t)$ represent the residual component; then:

$$y(t) = T(t) + S(t) + R(t). \tag{5.3}$$

In time-series decomposition, a commonly used method is STL (seasonal-trend decomposition using Loess), where the residual is obtained through local weighted regression (Loess). The formula for Loess is:

$$\hat{R}_t = y(t) - \hat{T}(t) - \hat{S}(t), \tag{5.4}$$

where $\hat{R}_t$ is the residual, $\hat{T}(t)$ is the estimated trend, and $\hat{S}(t)$ is the estimated seasonality. It finds utility in applications where anomalies are expected to manifest as deviations from expected patterns after accounting for trend and seasonality. Wu et al. propose a comprehensive analysis framework for anomaly detection in water supply systems, integrating time-series decomposition, outlier detection, and quantitative evaluation (Z. Y. Wu & He, 2021). Applied to a monitored water supply zone with over 8,000 pipes, the solution achieved a 75% true positive rate and successfully detected 90% of field events with a lead time of more than 24 hours, demonstrating its effectiveness in improving operational management and maximising the return of investment in smart water grids. It is commonly applied in domains such as finance, environmental monitoring, and energy consumption analysis. This approach allows for the separation of anomalous patterns from underlying trends and seasonal variations, enhancing interpretability. In Z. Zhang, Wang, Ding, and Gu (2024), authors utilise seasonal-trend decomposition to simplify analysis and improve detection performance, demonstrating state-of-the-art results across diverse anomalies. In addition, researchers have developed a decomposition approach to separate a time series into its latent component in time-series anomaly detection analysis, which offers control over type-I errors and robustness against anomalies (Choudhary, Hiranandani, & Saini, 2018). It offers robustness and can handle time-series data with complex temporal structures and multiple seasonalities. Additionally, it provides insights into the nature of anomalies by analysing residual patterns. However, time-series decomposition relies on the assumption that anomalies are captured in the residual component, which may not always hold true. It is sensitive to decomposition techniques, and performance may vary depending on the choice of method and parameters. Moreover, it requires additional computational resources compared to threshold-based methods due to decomposition processes.

5.3.1.3 Methods Based on Statistical Modeling

Statistical modeling assumes that time-series data follows a specific probability distribution, such as normal or exponential distribution. Anomalies are identified as data points that deviate significantly from the expected distribution according to the model. This method involves fitting a statistical model to the data and assessing the likelihood of observed data under the model. This approach is suitable for applications where anomalies are characterised by deviations from expected statistical properties, such as mean, variance, or distribution shape. Zeng et al. (Zeng, Yang, & Bo, 2020) propose a method for anomaly detection in wind turbine gearbox oil temperature, which utilises Sparse Bayesian Learning (SBL) and hypothesis testing (HT) to estimate variation ranges for different operating conditions and detect anomalies with high reliability. It also finds applications in quality control, network monitoring, and healthcare. Authors utilise denoising diffusion probabilistic models (DDPMs) for anomaly detection, leveraging a novel partial diffusion strategy named AnoDDPM, which outperforms existing methods by incorporating multi-scale simplex noise diffusion for improved anomaly detection (Wyatt, Leach, Schmon, & Willcocks, 2022). Statistical modeling provides a formal framework for quantifying the likelihood of anomalies based on probability theory. It offers adaptability, as it can accommodate various types of anomalies by selecting appropriate probability distributions or model parameters. Additionally, it is capable of detecting anomalies with diverse patterns and characteristics. However, the performance of statistical modeling heavily depends on the accuracy of the assumed probability distribution, which may not always reflect the true data-generating process. It entails complexity and requires expertise in statistical modeling and model selection, particularly for non-standard distributions or complex data patterns. Moreover, its performance may degrade in the presence of data quality issues, such as outliers or missing values in the dataset.

5.3.2 Traditional Machine Learning

Traditional machine learning methods for time-series anomaly detection typically involve extracting relevant features from the time-series data and training a model, such as a decision tree (DT), support vector machine (SVM), or neural network, to differentiate between normal and anomalous patterns based on these features. The underlying principle is to capture the distinctive characteristics of anomalies through feature-based representation and learn to distinguish them from normal patterns. These methods rely on the assumption that anomalies exhibit unique feature patterns that can be effectively learned by machine learning models.

5.3.2.1 Methods Based on Feature

The feature-based method involves extracting various features from time-series data and training traditional machine learning models using these features. These models differentiate between normal and abnormal patterns based on the extracted features. This approach is particularly useful in scenarios where anomalies exhibit distinct feature patterns that can be effectively captured and distinguished by machine learning models. Feature-based methods find wide applications across various domains, such as intrusion detection in cybersecurity (Cui, Wang, & Yue, 2019; Sarker, 2021; Yaseen, 2023) and fraud detection in financial transactions (Rtayli & Enneya, 2020). Furthermore, in industrial systems, Dhiman et al. introduced an anomaly detection model for wind turbine gearboxes based on a twin support vector machine (TWSVM), using SCADA data from wind farms in the UK (Dhiman, Deb, Muyeen, & Kamwa, 2021). These models are employed in situations where specific characteristics or patterns serve as indicators of anomalies. These methods offer flexibility in accommodating a wide range of feature types and extraction techniques, allowing for the interpretation of detected anomalies based on the importance of individual features. They can also generalise well to unseen data if the selected features adequately represent the underlying patterns. However, feature-based methods have their drawbacks. They heavily rely on domain knowledge and feature engineering expertise, making them sensitive to the quality and relevance of the selected features. Additionally, they may suffer from the curse of dimensionality when dealing with high-dimensional feature spaces, which can impact their performance.

5.3.2.2 Methods Based on Clustering

Clustering methods aim to group time-series data into different clusters and identify clusters that differ from others as anomalies. The principle behind this approach is that anomalies typically manifest as clusters that are distinct from the majority of the data. These methods find applications in scenarios where anomalies are characterised by unique cluster patterns, such as network traffic analysis, sensor data monitoring, and healthcare monitoring systems. Clustering methods are suitable for scenarios where anomalies can be identified based on their deviation from the majority of data points and are expected to form distinct clusters. Subudhi et al. (Subudhi & Panigrahi, 2020) introduce a novel approach for detecting fraud in automobile insurance claims using genetic algorithm-based fuzzy C-means clustering, combined with various supervised classifier models. The method effectively identifies genuine, malicious, and suspicious claims, leading to improved fraud detection in the automobile insurance domain. One advantage of clustering methods is their unsupervised nature, as they do not require labeled data for training, making them suitable for unsupervised anomaly detection scenarios. Ghezelbash et al. propose a genetic algorithm-based clustering method, known as genetic k-means clustering (GKMC) (Ghezelbash, Maghsoudi, & Carranza, 2020) to delineate multi-elemental patterns in stream sediment geochemical data for mineral exploration, demonstrating its effectiveness and reliability in geochemical anomaly recognition. They also offer flexibility in detecting anomalies with diverse patterns and characteristics without relying on predefined anomaly labels. Additionally, clustering methods can handle large volumes of time-series data efficiently. However, these methods have their limitations. They require careful selection of the number of clusters, which can be challenging and subjective. Performance may also vary based on the initialisation of cluster centroids and the choice of clustering algorithm. Furthermore, the interpretation of detected anomalies may be challenging due to the lack of explicit labels and the inherent complexity of cluster structures.

5.3.2.3 Methods Based on Distance

Distance-based methods identify anomalies by computing the distance or similarity measures between time-series data points, where anomalies are typically identified as data points that are significantly distant from others. The principle behind this approach is that anomalies often exhibit dissimilarities or distant relationships compared to normal data points. Let x_i and x_j represent two data points in the time-series data, and $\text{dist}(x_i, x_j)$ represent the distance between them. A common distance measure is the Euclidean distance:

$$\text{dist}(x_i, x_j) = \sqrt{\sum_{k=1}^{n} (x_{ik} - x_{jk})^2}, \tag{5.5}$$

where n is the number of dimensions/features. These methods find applications in scenarios where anomalies can be detected based on their dissimilarity or distance from normal patterns, such as anomaly detection in sensor networks, environmental monitoring, and outlier detection in financial data. Laskar et al. propose an intrusion detection system for industrial computer networks that combines the k-means algorithm with the Isolation Forest (Laskar et al., 2021). Distance-based methods are suitable for scenarios where anomalies can be identified based on their deviation from normal data points and are expected to exhibit distinct distance relationships. Henriques et al. introduce a distance-based anomaly detection model applied in the field of computing and networking systems, effectively processing large log file datasets to identify anomalous events, combining k-means and XGBoost models (Henriques, Caldeira, Cruz, & Simões, 2020). One advantage of distance-based methods is their intuitiveness, as anomalies are identified based on their distance from normal data points. They also offer flexibility, as they can be applied to various types of time-series data without restrictions on specific data distributions. Additionally, distance-based methods are robust and can handle noisy data and outliers effectively. However, these methods have their limitations. Performance heavily depends on the choice of distance metrics, which may vary based on the characteristics of the data. They may also suffer from the curse of dimensionality in high-dimensional feature spaces, leading to degraded performance. Furthermore, computing distances between data points can be computationally intensive, especially for large datasets and high-dimensional feature spaces.

5.3.3 Deep Learning

In the field of deep learning, there are various methods for time-series anomaly detection (Y. Liu et al., 2020). These methods can be categorised into six main types based on their core concepts and technical characteristics: RNN-based, CNN-based, autoencoder-based, GAN-based, attention-based, and GNN-based.

5.3.3.1 Methods Based on RNNs

RNNs are a class of neural network models capable of processing sequential data. Their structure includes recurrent connections, allowing the network to remember past information and apply it to current predictions. In time-series anomaly detection, RNNs can incrementally process each time step of sequential data and identify abnormal patterns by learning the temporal dependencies between data points (Ackerson, Dave, & Seliya, 2021; Murugesan & Thilagamani, 2020). Let x_t represent the t-th data point of the time series and h_t represent the hidden state of the RNN; then the propagation formula for RNN is:

$$h_t = f(x_t, h_{t-1}), \tag{5.6}$$

where f is the non-linear activation function of the RNN. RNNs are widely applied in scenarios requiring consideration of sequential data, such as natural language processing, speech recognition, and time-series prediction (X. Wei, Zhang, Yang, Zhang, & Yao, 2021). In anomaly detection, RNNs can be applied to various scenarios requiring capturing temporal dependencies, such as power system monitoring (Zhou & Musilek, 2021), network traffic analysis (G. Wei & Wang, 2021), and intelligent health monitoring (Dutta, Lenka, Nayak, Khandual, & Bhoi, 2021). In power system fault detection, RNNs can analyse power sensor data to identify abnormal power waveform patterns, such as voltage spikes or frequency anomalies (Madabhushi & Dewri, 2023). Additionally, in intelligent health monitoring, utilising RNNs can analyse patients' physiological signal data, such as heart rate, respiration rate, and body temperature, to detect abnormal physiological patterns like arrhythmias or sudden temperature rises. RNNs are suitable for processing sequential data with long-term dependencies and are sensitive to the temporal order of data. Due to their recurrent structure, RNNs can handle variable-length sequences and demonstrate excellent performance in modeling and predicting sequential data. A common variant of RNNs is LSTM, which captures long-term dependencies by introducing gating units. Another variant is gated recurrent units (GRUs), which incorporate gating mechanisms but with a simpler structure and fewer parameters. In time-series anomaly detection, these variant models often outperform traditional RNN models, as they better capture long-term dependencies in sequential data (H. Liu, Zhao, Wang, Yuan, & Feng, 2021). Research indicates that using variant RNN models like LSTM or GRU for time-series anomaly detection can effectively identify various types of abnormal patterns and generally outperform traditional RNN models in terms of performance and generalisation ability (Y. Wei, Jang-Jaccard, Xu, Sabrina, Camtepe, & Boulic, 2023). These methods can capture long-term dependencies in time-series data, handle sequences of different lengths, and process variable-length sequences. Moreover, RNN-based methods exhibit flexibility and generality in handling various types of time-series data. However, these methods also have some drawbacks, including difficulties in training with longer sequences, the problem of vanishing or exploding gradients, and the requirement for large amounts of data and computational resources.

5.3.3.2 Methods Based on CNNs

CNNs are deep learning models specifically designed for processing image data, but their application to time-series data involves capturing local features of sequential data. Through a combination of convolutional and pooling layers, CNNs can effectively extract local patterns and periodic variations in sequential data. CNNs are suitable for time-series anomaly detection scenarios requiring capturing local patterns or periodic variations, such as power system fault detection, sensor data anomaly detection, and speech recognition (Ullah, Ullah, Haq, Muhammad, Sajjad, & Baik, 2021). In sensor networks, CNNs can analyse sensor data streams to detect abnormal sensor measurement patterns, such as abnormal vibration patterns or temperature changes. Additionally, in network traffic analysis, CNNs can analyse network traffic data packets to detect abnormal network communication patterns, such as DDoS attacks or anomalous data transmission behaviour (Ghanbari & Kinsner, 2021). CNNs are suitable for cases where time-series data exhibit distinct local features or periodic patterns, and they are insensitive to the overall sequence order. A common variant of CNNs is one-dimensional convolutional neural networks (1D-CNN), specifically designed for processing sequential data. Another variant is CNN models with residual connections, such as ResNet, which can learn deeper features from data and often exhibit better performance and stability. Research suggests that using variant CNN models like 1D-CNN or ResNet for time-series anomaly detection can effectively capture local patterns and periodic variations in sequential data, thereby improving the accuracy and robustness of anomaly detection.

5.3.3.3 Methods Based on Autoencoders

Autoencoders are unsupervised learning models consisting of an encoder and a decoder, capable of encoding input data into a low-dimensional representation and attempting to decode it back to the original data. In time-series anomaly detection, the reconstruction error of autoencoders is used to identify abnormal patterns. Autoencoders are widely applied in scenarios requiring dimensionality reduction and reconstruction of data, such as image anomaly detection, financial fraud detection, and abnormal behaviour detection (C. Yin, Zhang, Wang, & Xiong, 2020). In financial fraud detection, autoencoders can model customer transaction data to identify abnormal transaction patterns, such as abnormal transaction amounts or frequencies. In manufacturing, autoencoders can analyse sensor data from production equipment to detect abnormal production process patterns, such as abnormal temperature changes or pressure fluctuations (L. Li, Yan, Wang, & Jin, 2020). Autoencoders are suitable for cases where time-series data have a potential low-dimensional representation and abnormal patterns differ significantly from normal patterns. Due to their unsupervised learning nature, autoencoders do not require labeled anomaly data for training, thus exhibiting a degree of flexibility and generality in anomaly detection. A common variant of autoencoders is variational autoencoders (VAE), which introduce latent variables to learn the underlying distribution of data and often have better generation and sampling capabilities (Longyuan, Junchi, Haiyang, & Yaohui, 2020). Another variant is sparse autoencoders, which learn more compact data representations by introducing sparsity constraints. Research indicates that using variant autoencoder models like VAE or sparse autoencoders for time-series anomaly detection can effectively capture the latent features of data and generally have better anomaly detection performance and generalisation ability.

5.3.3.4 Methods Based on GANs

Generative adversarial networks (GANs) consist of a generator and a discriminator. The generator attempts to generate data samples similar to the normal data distribution, while the discriminator tries to distinguish between generated fake samples and real samples. Through adversarial training, the generator gradually learns to generate samples closer to real data, while the discriminator becomes better at distinguishing between real and fake samples. Anomalies are considered data that deviates from the generated data. GANs are widely applied in fields such as image generation, image restoration, and style transfer. In time-series anomaly detection, they can be used to generate data samples similar to normal data, thereby detecting anomalous data (Geiger, Liu, Alnegheimish, Cuesta-Infante, & Veeramachaneni, 2020). A typical application is Anomaly Detection with Generative Adversarial Networks (AnoGAN) in time-series anomaly detection (Bashar & Nayak, 2020). AnoGAN learns the distribution of normal data by training the generator and discriminator to generate samples similar to normal data. During the generation process, anomalous data typically cannot generate samples similar to normal data, thereby enabling the identification of anomalies based on the output of the discriminator. These methods are suitable for cases where time-series data has complex data distributions and significant differences between abnormal and normal patterns. GAN-based models improve the performance and stability of GANs by introducing conditional information, improving loss functions, or implementing cross-domain data transformation. GANs-based models have achieved good results in some time-series anomaly detection tasks. They can effectively generate data samples similar to the normal data distribution and distinguish anomalous data. However, the training process of GANs is typically complex, requiring fine-tuning of hyperparameters and training techniques, and also demanding high-quality and feature-rich data.

5.3.3.5 Methods Based on Attention

Attention mechanisms allow models to dynamically focus on the most important parts when processing sequential data, thereby enhancing the model's ability to detect anomalies. Anomalies typically manifest as abnormal patterns that attract attention. Attention mechanisms are widely applied in natural language processing, speech recognition, bioinformatics, etc. In time-series anomaly detection, they can be used to capture key features in sequential data and improve anomaly detection accuracy (Javed, Usman, Rehman, Khan, & Haghighi, 2020). A common application is attention-based RNN models like the transformer model. The transformer model dynamically captures dependencies between different positions in sequence data through self-attention mechanisms, thereby enhancing the perception of anomalies (Tuli, Casale, & Jennings, 2022). These methods are suitable for cases where time-series data contain key features or patterns, and there are significant differences between abnormal and normal patterns. Variant attention mechanism models include multi-head attention mechanisms and variants of self-attention mechanisms. These variant models further enhance the performance and generalisation ability of models by introducing different attention mechanism designs. Models based on attention mechanisms have achieved significant results in tasks such as language modeling, sequence generation, and text classification. These can dynamically focus on important parts of sequence data, improving the accuracy of anomaly detection. However, these models typically require substantial computational resources and large-scale datasets for training and are sensitive to the selection of model structures and hyperparameters. These methods can dynamically focus on important parts of sequence data, improving the accuracy of anomaly detection. However, they require a large amount of training data and complex model structures and may suffer from overfitting, requiring a high level of understanding and modeling capability for sequence data.

5.3.3.6 Methods Based on GNNs

Graph neural networks can capture relationships and features between nodes in sequential data, identifying anomalous patterns by learning information from the graph structure. In time-series data, sequential data can be represented as a graph structure, where nodes represent data points and edges represent relationships or similarities between data points. Graph neural networks are widely applied in social network analysis, traffic flow analysis, bioinformatics, etc. (Deng & Hooi, 2021; Y. Wu, Dai, & Tang, 2021). In time-series anomaly detection, they can be used to capture node relationships and features in sequential data, thereby identifying anomalous patterns. A common application is the Graph Attention Network (GAT) based on graph neural networks. The GAT model effectively captures relationships and features between nodes in the graph structure through self-attention mechanisms, thus capturing abnormal patterns in sequential data (Zhao et al., 2020). These methods are suitable for cases where sequential data can be represented as a graph structure, and abnormal patterns typically manifest as anomalous relationships between nodes or edges. Variant graph neural network models include Graph Convolutional Networks (GCNs) (Atkinson, Bhardwaj, Englert, Ngairangbam, & Spannowsky, 2021), GraphSAGE (C. Chen, Li, Chen, Liang & Huang, 2023), and Deep Graph Infomax (Gao et al., 2023). These models improve the performance and robustness of the model by introducing different graph convolution operations, attention mechanism designs, or graph structure modeling methods. Models based on graph neural networks have achieved significant results in fields such as social network analysis, image segmentation, and bioinformatics (Z. Chen, Chen, Zhang, Yuan, & Cheng, 2021). These methods can effectively capture node relationships and features in sequential data, suitable for handling complex sequential data structures. However, they may require significant training and inference times for large graph structures and complex relationships, requiring a high level of understanding and processing capability for graph structures. Overall, these six categories of methods share the common goal of utilising deep learning techniques to achieve time-series anomaly detection. They employ different approaches and technical means to address various types of anomaly detection problems. Depending on factors such as data characteristics, task requirements, and model performance, one can choose a suitable method for conducting research and application in time-series anomaly detection.

5.4 APPLICATION

5.4.1 Smart Grid

The power grid is one of the most crucial infrastructures in modern society, and its stable operation is vital for ensuring social livelihoods and economic development. With the rapid development of smart grids and renewable energy sources, the scale and complexity of power grid data are continuously increasing. Power grid data exhibits high complexity and heterogeneity, including real-time monitoring data from sensors, measuring devices, and surveillance systems, as well as multidimensional time series covering aspects such as power grid status, load conditions, energy production, and consumption. This data is often influenced by various factors such as natural phenomena, human interference, and technical faults. Therefore, time-series anomaly detection has widespread applications in the field of power grids, including but not limited to abnormal load detection, voltage anomaly detection, fault diagnosis, and prediction. By promptly identifying and addressing abnormal situations, the reliability, safety, and economic efficiency of the power grid can be improved, ensuring stable and sustainable power supply. For example, Wei et al. introduce Grid Load Anomaly Detection (GLAD), a method for detecting load anomalies in microgrids using the enhanced ESD test (Q. Wei et al., 2020). GLAD shows promising results in simulating anomaly detection through statistical analyses. Expanding upon anomaly detection in power grids, Liang et al. present a fusion algorithm combining Isolation Forest and *k*-means to accurately detect abnormal access IP in power grid dispatching platforms (Shouyu et al., 2020). By addressing limitations of Isolation Forest, such as binary classification and threshold setting, the proposed approach demonstrates effectiveness and advancement in anomaly detection, as validated through evaluation metrics using data from the southern power grid dispatching platform. DYN-WATCH, a domain knowledge-based and topology-aware algorithm, effectively detects anomalies in power grid sensor data (S. Li, Pandey, Hooi, Faloutsos, & Pileggi, 2021). It accommodates regular topology adjustments, offering both accuracy and speed, with an average processing time of less than 1.7 ms for each sensor at every time step on a laptop computer, scaling linearly with the size of the graph. In addition, leveraging a CNN within the digital twin framework, William et al. propose a method for detecting physical faults in power systems (Danilczyk, Sun, & He, 2021). By analysing high-fidelity data from benchmark power systems, the approach not only detects faults but also classifies the affected bus. Oprea et al. (Oprea, Bâra, Puican, & Radu, 2021) explore the detection of errors and fraud in smart metering data using machine learning algorithms, aiming to alert utility companies to suspicious consumption behaviour. A spectral residual-convolutional neural network (SR-CNN) approach is proposed, which achieves a 90% accuracy, 0.875 precision score, and 0.894 F1 score in identifying suspicious consumers. Zhang et al. propose a self-supervised framework for anomaly detection in smart grid time-series data (Z. Zhang et al., 2022). The framework captures dynamic correlations between different sensors, effectively detecting anomalies in smart grid data. In Guha, Chatterjee, and Sikdar (2023), the authors present a deep generative model for anomaly detection in power grid data and apply it to transient stability assessment in the IEEE NewEngland-39 bus. Xiao et al. propose the Swin Transformer model for detecting abnormal power time-series data, enhancing prediction accuracy in power grid security (X. Xiao, Yang, & Gao, 2023). Park et al. introduce GridCAL, a context-aware anomaly detection algorithm for real-time power system SCADA data (Park & Pandey, 2024). By considering the impact of network topology and load/generation changes, GridCAL converts power flow measurements to context-agnostic values, enabling accurate anomaly detection across different grid contexts.

5.4.2 Network

In modern society, cybersecurity is paramount to safeguarding sensitive information and ensuring the uninterrupted operation of various digital systems. With the proliferation of networked devices and the increasing sophistication of cyber threats, the volume and complexity of cybersecurity data continue to grow exponentially. Cybersecurity data encompasses a wide range of heterogeneous sources, including network traffic logs, system event logs, user activity logs, and application logs, each capturing different aspects of system behaviour and potential security incidents. These data are susceptible to diverse threats such as malware infections, unauthorised access attempts, distributed denial-of-service (DDoS) attacks, and data breaches, among others. As a result, time-series anomaly detection plays a critical role in cybersecurity, offering a proactive approach to identify and mitigate emerging threats before they escalate into full-fledged cyberattacks. By analysing patterns and deviations in time-series data, anomaly detection algorithms can detect unusual behaviour indicative of potential security breaches or malicious activities. Common applications of time-series anomaly detection in cybersecurity include but are not limited to intrusion detection, network anomaly detection, malware detection, and fraud detection. By promptly identifying and responding to anomalies, organisations can enhance their cyber resilience, protect sensitive data, and maintain the integrity and availability of their digital assets. For instance, Goetz et al. propose an unsupervised anomaly detection approach for cyber-physical production systems using graph neural networks to model the system as a graph, incorporating correlations and interactions (Goetz & Humm, 2024). The introduced reconstruction-based graph neural network effectively detects anomalies in real industrial cyber-physical production systems. Yang et al. (Yang, Howley, & Schukat, 2024) introduce ADT, an agent-based dynamic thresholding method using deep reinforcement learning (DQN), for time-series anomaly detection in cyber-physical systems (CPS). Furthermore, Sun et al. propose MTS-DVGAN, an unsupervised dual variational generative adversarial model, for anomaly detection in multivariate time-series data in CPS (Sun, Huang, Han, Fu, Liu, & Long, 2024). Yin et al. introduce MSCBL-ADN, a novel model combining multi-scale CNN and bidirectional long-short term memory (bi-LSTM) arbitration dense network, for detecting low-rate distributed denial of service (LDDoS) attacks in cloud computing networks (X. Yin, Fang, Liu, & Liu, 2024). In Zavrak and Iskefiyeli (2023), the SAnDet (SDN anomaly detector) architecture is proposed as an anomaly-based intrusion detection system, leveraging software-defined networking (SDN) capabilities. The system utilises RNN and LSTM-based encoder-decoder (EncDecAD) methods to detect unknown attacks using flow features from OpenFlow switches. Kandhro et al. (2023) introduce a deep learning-based approach utilising a GAN for cybersecurity threat detection in IoT-driven cyber-physical systems, with superior performance in detecting various types of attacks. Thiruloga et al. introduce the TENET framework, which employs temporal convolutional neural networks with an integrated attention mechanism for anomaly detection in vehicles, addressing cybersecurity threats in modern vehicle systems (Thiruloga, Kukkala, & Pasricha, 2022).

5.4.3 Financial

In the realm of finance, time-series anomaly detection holds immense significance for ensuring the integrity and stability of financial markets and institutions. With the rapid evolution of financial technologies and the increasing complexity of trading systems, financial data has become voluminous and heterogeneous, comprising market prices, trading volumes, transaction records, and economic indicators, among others. These data streams are susceptible to various anomalies, including market manipulation, insider trading, fraudulent activities, and systemic risks, which can have far-reaching consequences on market stability and investor confidence. Time-series anomaly detection techniques play a pivotal role in identifying and mitigating such anomalies by analysing patterns and deviations in financial data. By leveraging statistical models, machine learning algorithms, and advanced data analytics techniques, financial institutions can detect irregularities in trading behaviour, detect fraudulent transactions, and identify emerging market trends or anomalies that may pose risks to financial stability. Common applications of time-series anomaly detection in finance include fraud detection in payment transactions, market surveillance for detecting abnormal trading activities, credit risk assessment, and algorithmic trading monitoring. By promptly detecting and responding to anomalies in financial time-series data, organisations can enhance market transparency, investor protection, and financial resilience, thereby fostering trust and confidence in financial markets and institutions. For examole, Aftabi et al. introduce a novel approach for fraud detection in financial statements using GAN and ensemble models (Aftabi, Ahmadi, & Farzi, 2023). It achieves competitive performance with supervised models and outperforming unsupervised methods. Khodabandehlou et al. (Khodabandehlou & Golpayegani, 2024) propose FiFrauD, an unsupervised and scalable approach for detecting fraudulent traders and suspicious behaviours in real-time financial transactions in e-markets. Furthermore, Chullamonthon et al. propose an ensemble model combining supervised LSTM network and unsupervised LSTM-based autoencoder deep learning techniques for detecting stock price manipulations (Chullamonthon & Tangamchit, 2023). Jiang et al. (Jiang, Dong, Wang, & Xia, 2023) propose the UAAD-FDNet framework for credit card fraud detection, applying unsupervised attentional anomaly detection networks. It is designed to detect fraudulent transactions from large-scale transaction datasets and shows good performance. Habibpour et al. propose three uncertainty quantification (UQ) techniques—Monte Carlo dropout, ensemble, and ensemble Monte Carlo dropout—for credit card fraud detection using transaction data (Habibpour et al., 2023). Yan et al. (Yan et al., 2020) introduce the CEEMD-PCA-LSTM model for stock market anomalies detection, combining complementary ensemble empirical mode decomposition (CEEMD), principal component analysis (PCA), and LSTM networks.

5.5 CHALLENGES AND FUTURE WORKS

Despite the wide-ranging applications of time-series anomaly detection, significant challenges persist in its implementation and deployment.

5.5.1 Model Interpretability

Many time-series anomaly detection models adopt complex deep learning structures such as RNNs, LSTMs, and CNNs. These models possess a large number of parameters and layers, making it difficult to intuitively understand the inner workings and decision-making processes of the models. Additionally, deep learning-based time-series anomaly detection models are often perceived as black-box models, where their internal logic and decision-making processes are invisible or opaque to the user. This lack of transparency prevents users from understanding how the model processes data and detects anomalies, leading to a lack of trust in practical applications and making the model results difficult to interpret and validate. Future research can focus on the following methods to improve model interpretability:

- **Model simplification**: Enhance model interpretability by designing simplified model structures or adding specialised interpretability components. Attention mechanisms allow models to focus on the most relevant parts of the data, making the decision-making process more understandable. For example, attention mechanisms can be used in time-series anomaly detection to highlight anomalous data points or patterns. Interpretability layers are special layers used to transform the model's output into a more understandable form, such as probability distributions or interpretable rules. Rule engines allow domain expert knowledge to be directly integrated into the model to form rule-based interpretable models.
- **Construction of interpretable models**: Developing specialised interpretable time-series anomaly detection models is an effective method to improve model interpretability. Interpretable models are typically based on simple algorithms or rules, such as rule-based methods, decision trees, or linear models. Compared to complex deep learning models, these models are easier to understand and interpret because their decision-making processes are more transparent. By constructing specialised interpretable models, users can more intuitively understand anomaly detection results and comprehend the workings and decision logic of the model.

5.5.2 Large-Scale and High-Dimensional Data

With the proliferation of the internet and the development of IoT, an increasing amount of data is being generated and collected, leading to exponential growth in data volume. Additionally, data in many fields not only have a large number of samples but also may have a large number of features, resulting in increased data dimensionality. For example, data collected in sensor networks may have thousands or even millions of features, making traditional anomaly detection algorithms ineffective. Future research can focus on utilising the statistical properties of data and redundancy information to address the challenges of large-scale and high-dimensional data processing.

- **Distributed computing**: Distributed computing processes data distributed across multiple computing nodes, significantly reducing computational complexity and processing time. By employing parallel computing and distributed algorithms, large-scale datasets can be divided into multiple small batches, each processed on different computing nodes, and then the results are merged. This allows for effective utilisation of computational resources and handling datasets larger than the memory capacity of individual computing nodes.
- **Compression and dimensionality reduction**: Compression and dimensionality reduction techniques reduce data redundancy and feature dimensionality, greatly reducing storage requirements and computational overhead. For example, compression methods based on sampling can reduce the number of data samples through random sampling or importance sampling, thereby reducing storage requirements. Dimensionality reduction techniques such as principal component analysis (PCA) or singular value decomposition (SVD) project high-dimensional data into low-dimensional subspaces, reducing the number of features and computational complexity. Therefore, future developments will focus on developing efficient data processing techniques, including algorithms based on distributed and parallel computing, data compression and dimensionality reduction techniques, and GPU acceleration. Additionally, for high-dimensional data, attention should be focused on feature selection and feature extraction techniques to reduce data dimensionality and improve algorithm efficiency.

5.5.3 Model Lifelong Learning

Traditional anomaly detection models are often trained on static or representative training datasets, which may not fully cover various situations and changes in the real world. When the model encounters data distributions different from the training data, its performance may significantly deteriorate because the model cannot learn sufficiently generalised representations or patterns from the training data. Additionally, model parameters are often fixed and cannot be dynamically adjusted based on data changes. This means that once the model is trained, its behaviour remains unchanged in subsequent applications, and it cannot adapt flexibly to changes in data distribution or concept drift. Future research can focus on developing anomaly detection models with continual learning capabilities to address changes in data distribution and concept drift:

- **Online learning**: Utilising online learning techniques, models can continuously learn from new data and adjust model parameters in a timely manner based on data changes. By updating the model in real-time, it can adapt to changes in data distribution, thereby maintaining model performance and accuracy.
- **Transfer learning**: Leveraging transfer learning principles, previously learned knowledge and models can be transferred to new data domains, accelerating the learning process for new data. Through transfer learning, models can leverage existing knowledge and experience to quickly adapt to the characteristics and distribution of new data, improving model generalisation and adaptability.
- **Incremental learning**: Adopting incremental learning techniques allows models to continuously learn from new data and gradually accumulate new knowledge and experience. Through incremental learning, models can continuously update their parameters or structures to adapt to changes in data distribution and concept drift, thereby maintaining model performance and robustness.

5.5.4 Data Imbalance and Class Skewness

In many practical scenarios, the normal time-series data often exhibits a higher frequency than anomalous data. For instance, in monitoring systems, the normal operational state typically dominates the majority of time, with anomalous events being sporadic occurrences. This inherent data distribution imbalance results in issues of data imbalance and class skewness. In real-world applications, normal data instances are usually far more abundant than anomalous ones, leading to problems of data imbalance and class skewness. Future research can concentrate on the following methodologies to develop anomaly detection algorithms tailored to imbalanced data:

- **Cost-sensitive learning**: This methodology involves assigning varying weights or costs to samples from different classes during model training to mitigate the data imbalance problem. By adjusting the loss function, the model imposes greater penalties for misclassifying anomalous samples. The weights assigned to normal and anomalous samples in the loss function can be set differently, allowing the model to focus more on the accurate classification of anomalous samples during prediction. Additionally, distinct cost weights can be assigned to samples of each class during training, enabling the model to prioritise classes that are challenging to classify, thereby enhancing performance in imbalanced environments.
- **Meta-learning:** Meta-learning is an approach that enhances model performance and generalisation by learning the process of learning itself. In imbalanced environments, meta-learning can improve performance by acquiring the ability to adapt to imbalanced data distributions. By incorporating meta-learning algorithms during model training, the model can dynamically adjust its learning strategy to accommodate the characteristics of imbalanced data. By learning how to design loss functions suitable for imbalanced data, the model can effectively handle anomalous samples and enhance performance in imbalanced environments.

5.5.5 Multi-Modal Data Fusion

Time-series data may originate from various data sources and sensors, such as text descriptions, images, and sensor measurements. Each data source possesses its unique data modality and feature representation, resulting in the generation of multi-modal data. With the extensive adoption of multi-modal data across various domains, future research can explore novel methods of multi-modal data fusion to fully exploit the correlation and complementarity among different data sources.

- **Large language models (LLMs)**: LLMs, such as GPT and BERT, acquire language representations by pre-training on extensive text corpora, which contain abundant semantic and syntactic information. In multi-modal fusion, the textual representation capability of LLMs can be harnessed to fuse text data with other modalities. LLMs can transform textual data into high-dimensional semantic vector representations, offering a unified semantic space that facilitates semantic correlation and fusion across different data modalities.
- **Knowledge graphs:** Knowledge graphs are graphical structures employed to represent and organise knowledge, containing rich entity and relationship information. In multi-modal fusion, the structured representation capability of knowledge graphs can be leveraged to map data from different modalities onto the knowledge graph and exploit the entities and relationships therein for data correlation and fusion. Knowledge graphs provide semantically rich relationship information, aiding the model in better understanding and analysing multi-modal data.

5.6 CONCLUSION

In conclusion, the detection of anomalies in IoT time-series data is of paramount importance for ensuring the reliability, security, and efficiency of IoT systems across various domains. This chapter has provided a comprehensive overview of anomaly detection techniques in IoT, encompassing statistical methods, machine learning algorithms, and deep learning models for point anomalies, collective anomalies, and contextual anomalies. Despite the progress made in this field, several challenges such as data heterogeneity, scalability, interpretability, and real-time processing persist. Future research efforts should focus on addressing these challenges and exploring innovative approaches to enhance the effectiveness and efficiency of anomaly detection in IoT time-series data. By overcoming these challenges and advancing the state-of-the-art techniques, we can unlock the full potential of IoT applications and facilitate the realisation of a smarter and more connected world.

FUNDING

This work was supported by the National Natural Science Foundation of China (62072360, 62172438), the key research and development plan of Shaanxi province (2021ZDLGY02-09, 2023-GHZD-44, 2023-ZDLGY-54), the Natural Science Foundation of Guangdong Province of China (2022A1515010988), Key Project on Artificial Intelligence of Xi'an Science and Technology Plan (23ZDCYJSGG0021-2022, 23ZDCYYYCJ0008, 23ZDCYJSGG0002-2023), and the Proof-of-Concept Fund from Hangzhou Research Institute of Xidian University (GNYZ2023QC0201).

REFERENCES

1. Ackerson, J. M., Dave, R., & Seliya, N. (2021). Applications of recurrent neural network for biometric authentication & anomaly detection. *Information*, 12(7), 272.
2. Aftabi, S. Z., Ahmadi, A., & Farzi, S. (2023). Fraud detection in financial statements using data mining and GAN models. *Expert Systems with Applications*, 227, 120144.
3. Apostol, E. S., Truică, C. O., Pop, F., & Esposito, C. (2021). Change point enhanced anomaly detection for IoT time series data. *Water*, 13(12), 1633.
4. Atkinson, O., Bhardwaj, A., Englert, C., Ngairangbam, V. S., & Spannowsky, M. (2021). Anomaly detection with convolutional graph neural networks. *Journal of High Energy Physics*, 2021(8), 1–19.
5. Bashar, M. A., & Nayak, R. (2020). Tanogan: Time series anomaly detection with generative adversarial networks. In *2020 IEEE Symposium Series on Computational Intelligence (SSCI)* (pp. 1778–1785). IEEE.
6. Beteto, A., Melo, V., Lin, J., Alsultan, M., Dias, E. M., Korte, E., … Lambert, J. H. (2022). Anomaly and cyber fraud detection in pipelines and supply chains for liquid fuels. *Environment Systems and Decisions*, 42(2), 306–324.
7. Chen, C., Li, Q., Chen, L., Liang, Y., & Huang, H. (2023). An improved graphSAGE to detect power system anomaly based on time-neighbor feature. *Energy Reports*, 9, 930–937.

8. Chen, Z., Chen, D., Zhang, X., Yuan, Z., & Cheng, X. (2021). Learning graph structures with transformer for multivariate time-series anomaly detection in IoT. *IEEE Internet of Things Journal*, 9(12), 9179–9189.
9. Choudhary, S., Hiranandani, G., & Saini, S. K. (2018). Sparse decomposition for time series forecasting and anomaly detection. In *Proceedings of the 2018 SIAM International Conference on Data Mining* (pp. 522–530). IEEE.
10. Chullamonthon, P., & Tangamchit, P. (2023). Ensemble of supervised and unsupervised deep neural networks for stock price manipulation detection. *Expert Systems with Applications*, 220, 119698.
11. Cook, A. A., Mısırlı, G., & Fan, Z. (2019). Anomaly detection for IoT time-series data: A survey. *IEEE Internet of Things Journal*, 7(7), 6481–6494.
12. Cui, M., Wang, J., & Yue, M. (2019). Machine learning-based anomaly detection for load forecasting under cyberattacks. *IEEE Transactions on Smart Grid*, 10(5), 5724–5734.
13. Danilczyk, W., Sun, Y. L., & He, H. (2021). Smart grid anomaly detection using a deep learning digital twin. In *2020 52nd North American Power Symposium (NAPS)* (pp. 1–6). IEEE.
14. David Akande, T., Kaur, B., Dadkhah, S., & Ghorbani, A. A. (2022). Threshold based technique to detect anomalies using log files. In *Proceedings of the 2022 7th International Conference on Machine Learning Technologies* (pp. 191–198). ACM.
15. Deng, A., & Hooi, B. (2021). Graph neural network-based anomaly detection in multivariate time series. In *Proceedings of the AAAI Conference on Artificial Intelligence* (Vol. 35, pp. 4027–4035). Part of the PKP Publishing Services Network.
16. Dhiman, H. S., Deb, D., Muyeen, S., & Kamwa, I. (2021). Wind turbine gearbox anomaly detection based on adaptive threshold and twin support vector machines. *IEEE Transactions on Energy Conversion*, 36(4), 3462–3469.
17. Dutta, K., Lenka, R., Nayak, S. R., Khandual, A., & Bhoi, A. K. (2021). Med-net: A novel approach to ECG anomaly detection using LSTM auto-encoders. *International Journal of Computer Applications in Technology*, 65(4), 343–357.

18. Gao, P., Zhang, H., Wang, M., Yang, W., Wei, X., Lv, Z., & Ma, Z. (2023). Deep temporal graph infomax for imbalanced insider threat detection. *Journal of Computer Information Systems*, 65(1), 108–118.
19. Garg, A., Zhang, W., Samaran, J., Savitha, R., & Foo, C. S. (2021). An evaluation of anomaly detection and diagnosis in multivariate time series. *IEEE Transactions on Neural Networks and Learning Systems*, 33(6), 2508–2517.
20. Geiger, A., Liu, D., Alnegheimish, S., Cuesta-Infante, A., & Veeramachaneni, K. (2020). Tadgan: Time series anomaly detection using generative adversarial networks. In *2020 IEEE International Conference on Big Data (Big Data)* (pp. 33–43). IEEE.
21. Ghanbari, M., & Kinsner, W. (2021). Detecting DDoS attacks using an adaptive-wavelet convolutional neural network. In *2021 IEEE Canadian Conference on Electrical and Computer Engineering (CCECE)* (pp. 1–7). IEEE.
22. Ghezelbash, R., Maghsoudi, A., & Carranza, E. J. M. (2020). Optimization of geochemical anomaly detection using a novel genetic k-means clustering (GKMC) algorithm. *Computers & Geosciences*, 134, 104335.
23. Goetz, C., & Humm, B. G. (2024). Unsupervised correlation- and interaction-aware anomaly detection for cyber-physical production systems based on graph neural networks. *Procedia Computer Science*, 232, 2057–2071.
24. Guha, D., Chatterjee, R., & Sikdar, B. (2023). Anomaly detection using LSTM-based variational autoencoder in unsupervised data in power grid. *IEEE Systems Journal*, 17(3), 4313–4323.
25. Habibpour, M., Gharoun, H., Mehdipour, M., Tajally, A., Asgharnezhad, H., Shamsi, A., … Nahavandi, S. (2023). Uncertainty-aware credit card fraud detection using deep learning. *Engineering Applications of Artificial Intelligence*, 123, 106248.
26. Hao, W., Yang, T., & Yang, Q. (2021). Hybrid statistical-machine learning for real-time anomaly detection in industrial cyber–physical systems. *IEEE Transactions on Automation Science and Engineering*, 20(1), 32–46.
27. Hemdan, E. E. D., & Manjaiah, D. (2022). Anomaly credit card fraud detection using deep learning. *Deep Learning in Data Analytics: Recent Techniques, Practices and Applications*, 207–217.

28. Henriques, J., Caldeira, F., Cruz, T., & Simões, P. (2020). Combining k-means and XGBoost models for anomaly detection using log datasets. *Electronics*, 9(7), 1164.
29. Huong, T. T., Bac, T. P., Long, D. M., Luong, T. D., Dan, N. M., Thang, B. D., ... others (2021). Detecting cyberattacks using anomaly detection in industrial control systems: A federated learning approach. *Computers in Industry*, 132, 103509.
30. Javed, A. R., Usman, M., Rehman, S. U., Khan, M. U., & Haghighi, M. S. (2020). Anomaly detection in automated vehicles using multistage attention-based convolutional neural network. *IEEE Transactions on Intelligent Transportation Systems*, 22(7), 4291–4300.
31. Jiang, S., Dong, R., Wang, J., & Xia, M. (2023). Credit card fraud detection based on unsupervised attentional anomaly detection network. *Systems*, 11(6), 305.
32. Kandhro, I. A., Alanazi, S. M., Ali, F., Kehar, A., Fatima, K., Uddin, M., & Karuppayah, S. (2023). Detection of real-time malicious intrusions and attacks in IoT empowered cybersecurity infrastructures. *IEEE Access*, 11, 9136–9148.
33. Khodabandehlou, S., & Golpayegani, A. H. (2024). Fifraud: Unsupervised financial fraud detection in dynamic graph streams. *ACM Transactions on Knowledge Discovery from Data*, 18(5), 1–29.
34. Kong, F., Li, J., Jiang, B., Wang, H., & Song, H. (2021). Integrated generative model for industrial anomaly detection via bidirectional LSTM and attention mechanism. *IEEE Transactions on Industrial Informatics*, 19(1), 541–550.
35. Laskar, M. T. R., Huang, J. X., Smetana, V., Stewart, C., Pouw, K., An, A., ... Liu, L. (2021). Extending isolation forest for anomaly detection in big data via k-means. *ACM Transactions on Cyber-Physical Systems (TCPS)*, 5(4), 1–26.
36. Li, G., & Jung, J. J. (2023). Deep learning for anomaly detection in multivariate time series: Approaches, applications, and challenges. *Information Fusion*, 91, 93–102.
37. Li, J., Izakian, H., Pedrycz, W., & Jamal, I. (2021). Clustering-based anomaly detection in multivariate time series data. *Applied Soft Computing*, 100, 106919.
38. Li, L., Yan, J., Wang, H., & Jin, Y. (2020). Anomaly detection of time series with smoothness-inducing sequential variational

auto-encoder. *IEEE Transactions on Neural Networks and Learning Systems*, 32(3), 1177–1191.

39. Li, S., Pandey, A., Hooi, B., Faloutsos, C., & Pileggi, L. (2021). Dynamic graph-based anomaly detection in the electrical grid. *IEEE Transactions on Power Systems*, 37(5), 3408–3422.
40. Liu, H., Zhao, H., Wang, J., Yuan, S., & Feng, W. (2021). LSTM-GAN-AE: A promising approach for fault diagnosis in machine health monitoring. *IEEE Transactions on Instrumentation and Measurement*, 71, 1–13.
41. Liu, Y., Garg, S., Nie, J., Zhang, Y., Xiong, Z., Kang, J., & Hossain, M. S. (2020). Deep anomaly detection for time-series data in industrial IoT: A communication-efficient on-device federated learning approach. *IEEE Internet of Things Journal*, 8(8), 6348–6358.
42. Longyuan, L., Junchi, Y., Haiyang, W., & Yaohui, J. (2020). Anomaly detection of time series with smoothness-inducing sequential variational auto-encoder. *IEEE Transactions on Neural Networks and Learning Systems*, 32(3), 1177–1191.
43. Madabhushi, S., & Dewri, R. (2023). A survey of anomaly detection methods for power grids. *International Journal of Information Security*, 22(6), 1799–1832.
44. Mestav, K. R., Wang, X., & Tong, L. (2022). A deep learning approach to anomaly sequence detection for high-resolution monitoring of power systems. *IEEE Transactions on Power Systems*, 38(1), 4–13.
45. Moschini, G., Houssou, R., Bovay, J., & Robert-Nicoud, S. (2021). Anomaly and fraud detection in credit card transactions using the ARIMA model. *Engineering Proceedings*, 5(1), 56.
46. Mothukuri, V., Khare, P., Parizi, R. M., Pouriyeh, S., Dehghantanha, A., & Srivastava, G. (2021). Federated-learning-based anomaly detection for IoT security attacks. *IEEE Internet of Things Journal*, 9(4), 2545–2554.
47. Murugesan, M., & Thilagamani, S. (2020). Efficient anomaly detection in surveillance videos based on multi layer perception recurrent neural network. *Microprocessors and Microsystems*, 79, 103303.
48. Oprea, S. V., Bâra, A., Puican, F. C., & Radu, I. C. (2021). Anomaly detection with machine learning algorithms and big data in electricity consumption. *Sustainability*, 13(19), 10963.
49. Pang, G., Cao, L., & Aggarwal, C. (2021). Deep learning for anomaly detection: Challenges, methods, and opportunities. In *Proceedings*

of the 14th ACM International Conference on Web Search and Data Mining (pp. 1127–1130). ACM.

50. Park, S., & Pandey, A. (2024). Anomaly detection in power grids via context-agnostic learning. *arXiv preprint arXiv:2404.07898.*
51. Pérez, S. I., Moral-Rubio, S., & Criado, R. (2021). A new approach to combine multiplex networks and time series attributes: Building intrusion detection systems (IDS) in cybersecurity. *Chaos, Solitons & Fractals*, 150, 111143.
52. Rashid, A. B., Ahmed, M., Sikos, L. F., & Haskell-Dowland, P. (2022). Anomaly detection in cybersecurity datasets via cooperative co-evolution-based feature selection. *ACM Transactions on Management Information Systems (TMIS)*, 13(3), 1–39.
53. Rtayli, N., & Enneya, N. (2020). Selection features and support vector machine for credit card risk identification. *Procedia Manufacturing*, 46, 941–948.
54. Sarker, I. H. (2021). Cyberlearning: Effectiveness analysis of machine learning security modeling to detect cyber-anomalies and multi-attacks. *Internet of Things*, 14, 100393.
55. Shouyu, L., Zhang, K., Fang, W., Zhou, Z., Hu, R., Zhu, W., … Hou, J. (2020). Anomaly detection of power grid dispatching platform based on isolation forest and k-means fusion algorithm. In *Journal of physics: Conference series* (Vol. 1601, p. 022010). Iopscience Press.
56. Subudhi, S., & Panigrahi, S. (2020). Use of optimized fuzzy c-means clustering and supervised classifiers for automobile insurance fraud detection. *Journal of King Saud University-Computer and Information Sciences*, 32(5), 568–575.
57. Sun, H., Huang, Y., Han, L., Fu, C., Liu, H., & Long, X. (2024). MTS-DVGAN: Anomaly detection in cyber-physical systems using a dual variational generative adversarial network. *Computers & Security*, 139, 103570.
58. Thiruloga, S. V., Kukkala, V. K., & Pasricha, S. (2022). TENET: Temporal CNN with attention for anomaly detection in automotive cyber-physical systems. In *2022 27th Asia and South Pacific Design Automation Conference (ASP-DAC)* (pp. 326–331). IEEE.
59. Tuli, S., Casale, G., & Jennings, N. R. (2022). TranAD: Deep transformer networks for anomaly detection in multivariate time series data. *arXiv preprint arXiv:2201.07284.*

60. Ullah, W., Ullah, A., Haq, I. U., Muhammad, K., Sajjad, M., & Baik, S. W. (2021). CNN features with bi-directional LSTM for real-time anomaly detection in surveillance networks. *Multimedia Tools and Applications*, 80, 16979–16995.
61. Varone, G., Boulila, W., Lo Giudice, M., Benjdira, B., Mammone, N., Ieracitano, C., … others (2021). A machine learning approach involving functional connectivity features to classify rest-EEG psychogenic non-epileptic seizures from healthy controls. *Sensors*, 22(1), 129.
62. Vos, K., Peng, Z., Jenkins, C., Shahriar, M. R., Borghesani, P., & Wang, W. (2022). Vibration-based anomaly detection using LSTM/SVM approaches. *Mechanical Systems and Signal Processing*, 169, 108752.
63. Wei, G., & Wang, Z. (2021). Adoption and realization of deep learning in network traffic anomaly detection device design. *Soft Computing*, 25(2), 1147–1158.
64. Wei, Q., Ma, R., Wang, Y., Chen, M., Sun, Y., Liu, M., & Lin, X. (2020). GLAD: A method of microgrid anomaly detection based on ESD in smart power grid. In *2020 IEEE International Conference on Power, Intelligent Computing and Systems (ICPICS)* (pp. 103–107). IEEE.
65. Wei, X., Zhang, L., Yang, H. Q., Zhang, L., & Yao, Y. P. (2021). Machine learning for pore-water pressure time-series prediction: Application of recurrent neural networks. *Geoscience Frontiers*, 12(1), 453–467.
66. Wei, Y., Jang-Jaccard, J., Xu, W., Sabrina, F., Camtepe, S., & Boulic, M. (2023). LSTM-autoencoder-based anomaly detection for indoor air quality time-series data. *IEEE Sensors Journal*, 23(4), 3787–3800.
67. Wolleb, J., Bieder, F., Sandkühler, R., & Cattin, P. C. (2022). Diffusion models for medical anomaly detection. In *International Conference on Medical Image Computing and Computer-Assisted Intervention* (pp. 35–45). Springer.
68. Wu, Y., Dai, H. N., & Tang, H. (2021). Graph neural networks for anomaly detection in industrial internet of things. *IEEE Internet of Things Journal*, 9(12), 9214–9231.
69. Wu, Z. Y., & He, Y. (2021). Time series data decomposition-based anomaly detection and evaluation framework for operational management of smart water grid. *Journal of Water Resources Planning and Management*, 147(9), 04021059.

70. Wyatt, J., Leach, A., Schmon, S. M., & Willcocks, C. G. (2022). AnoDDPM: Anomaly detection with denoising diffusion probabilistic models using simplex noise. In *Proceedings of the IEEE/CVF Conference on Computer Vision and Pattern Recognition* (pp. 650–656). IEEE.
71. Xiao, X., Yang, Z., & Gao, X. (2023). Anomaly detection of power time-series data based on multi-dimensional transformer network. *Computer-Aided Design and Applications*, 15, 1–13.
72. Xiao, Z., & Jiao, J. (2021). Explainable fraud detection for few labeled time series data. *Security and Communication Networks*, 2021, 1–9.
73. Yan, B., Aasma, M., et al. (2020). A novel deep learning framework: Prediction and analysis of financial time series using CEEMD and LSTM. *Expert Systems with Applications*, 159, 113609.
74. Yang, X., Howley, E., & Schukat, M. (2024). ADT: Time series anomaly detection for cyber-physical systems via deep reinforcement learning. *Computers & Security*, 141, 103825.
75. Yaseen, A. (2023). The role of machine learning in network anomaly detection for cybersecurity. *Sage Science Review of Applied Machine Learning*, 6(8), 16–34.
76. Yin, C., Zhang, S., Wang, J., & Xiong, N. N. (2020). Anomaly detection based on convolutional recurrent autoencoder for IoT time series. *IEEE Transactions on Systems, Man, and Cybernetics: Systems*, 52(1), 112–122.
77. Yin, X., Fang, W., Liu, Z., & Liu, D. (2024). A novel multi-scale CNN and Bi-LSTM arbitration dense network model for low-rate DDoS attack detection. *Scientific Reports*, 14(1), 5111.
78. Zavrak, S., & Iskefiyeli, M. (2023). Flow-based intrusion detection on software-defined networks: a multivariate time series anomaly detection approach. *Neural Computing and Applications*, 35(16), 12175–12193.
79. Zeng, X., Yang, M., & Bo, Y. (2020). Gearbox oil temperature anomaly detection for wind turbine based on sparse Bayesian probability estimation. *International Journal of Electrical Power & Energy Systems*, 123, 106233.
80. Zhang, J. E., Wu, D., & Boulet, B. (2021). Time series anomaly detection for smart grids: A survey. In *2021 IEEE Electrical Power and Energy Conference (EPEC)* (pp. 125–130). IEEE.

81. Zhang, Y., Chen, Y., Wang, J., & Pan, Z. (2021). Unsupervised deep anomaly detection for multi-sensor time-series signals. *IEEE Transactions on Knowledge and Data Engineering*, 35(2), 2118–2132.
82. Zhang, Z., Wang, R., Ding, R., & Gu, Y. (2024). Unravel anomalies: An end-to-end seasonal-trend decomposition approach for time series anomaly detection. In *LCASSP 2024–2024 IEEE International Conference on Acoustics, Speech and Signal Processing (ICASSP)* (pp. 5415–5419). IEEE.
83. Zhang, Z., Zhao, L., Cai, D., Feng, S., Miao, J., Guan, Y., … Cao, J. (2022). Time series anomaly detection for smart grids via multiple self-supervised tasks learning. In *2022 IEEE International Conference on Knowledge Graph (ICKG)* (pp. 392–397).
84. Zhao, H., Wang, Y., Duan, J., Huang, C., Cao, D., Tong, Y., … Zhang, Q. (2020). Multivariate time-series anomaly detection via graph attention network. In *2020 IEEE International Conference on Data Mining (ICDM)* (pp. 841–850). IEEE.
85. Zhou, M., & Musilek, P. (2021). Real-time anomaly detection in distribution grids using long short term memory network. In *2021 IEEE Electrical Power and Energy Conference (EPEC)* (pp. 208–213). IEEE.

CHAPTER 6

Pen Ink Analysis in Handwritten Document Forensics

Classification and Current Trends

Rojalin Dash, Prabhat Dansena, Prasun Chandra Tripathi, and Ashish Ranjan

6.1 INTRODUCTION

As human beings have evolved, verbal and non-verbal communication has played an important role in the conveyance of emotions. In the case of non-verbal communication, people often write what they wish to be conveyed on paper or another medium. In this way, handwriting evolved throughout the world. Sometimes, to gain a financial or personal benefit, written documents are tampered with or altered. The alteration of any handwritten document is called handwritten document forgery. This type of unethical practice can cause financial and social loss to society. According to Ubelaker (2019), handwritten documents and signatures are considered the primary source of evidence across the globe. This mostly includes cases of forgery in cheques, credit card receipts, bomb threats, wills, or some types of financial dispute. It includes forgery in handwritten documents, scanned documents by a normal scanner, and documents produced by a computer or printer. So, to check the authenticity of these documents, they should be properly examined by a professional examiner. Document examiners normally give a rating scale indicating the relevance of the document according to the case (Adam, 2016).

DOI: 10.1201/9781003463559-6

TABLE 6.1 Classification of handwritten document forgery

Type	Writer	Pen	Action
1	Different	Same pen	Addition of new letter/word with original text
2	Same	Same pen	Alteration of new letter/word with existing text
3	Different	Different pen with similar colour	Addition of new letter/word with original existing text

As discussed in Koppenhaver (2007) and Ubelaker (2019) regarding the basic principles of handwriting, two people do not have the same set of handwriting, and one person cannot write or sign in a similar way every time; it varies due to physical, psychological, and circumstantial factors. Based on this, unethical practice in written documents is divided into three types, as depicted in Table 6.1.

6.1.1 Classification of Handwritten Document Forgery

This section delineates the types of handwritten document forgeries. Handwritten document forgery can be classified into three categories, as depicted in Table 6.1, that are discussed in the following subsections.

6.1.1.1 Type 1: Addition of New Word by Different Writer and Same Pen

There are some cases in crimes where some new word or letter is adjusted from the existing original word or text by a different writer. This alteration could change a monetary amount or ownership of the document, which is considered a serious offence. Moreover, identification of such alterations can be difficult, as similar ink may have been used. Example: In a case mentioned in Wadhwa, Maheshwari, Dansena, and Bag (2018), the payee name on a cheque is Smita Kumari. But the fraudster adds 'M.' to the existing name, which refers to some other person. In the amount section, the digit 7 is added to the existing digit 545, increasing the payable amount, as depicted Figure 6.1.

6.1.1.2 Type 2: Alteration of Word or Letter by Same Person and Similar Colour Pen

Alteration is changing a word or digit to a similar type of appearance. When someone is issuing a cheque or a financial document, he/she is careful to

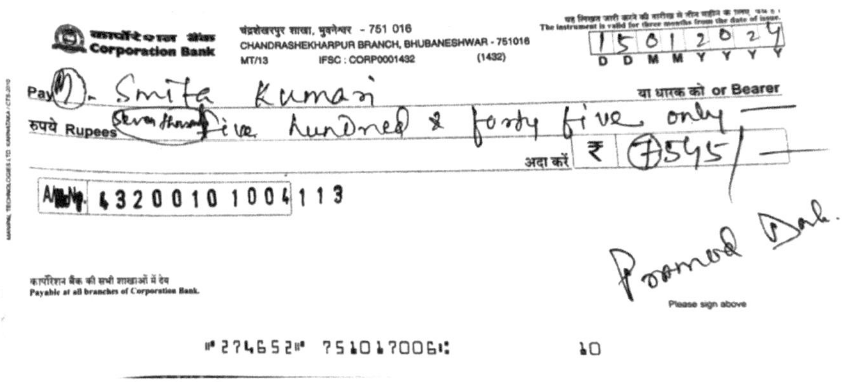

FIGURE 6.1 Addition of new words by the same writer and same pen.

mention the amount, name, date, and age. An unethical person with access to this document may covertly alter the digit by adding one line or a figure. This is also considered a case of forgery.

As per Dansena, Kumar, and Pal (2015), Indian cheques are carefully investigated, and important features are extracted by applying skewing techniques. Cheques contain important fields such as 'Payee Name,' 'Courtesy Amount,' 'Legal Amount,' 'Date,' and 'Account Number.' Questioned documents are scanned and converted into binary value using Otsu's method, and important edges are detected by following Canny edge detection techniques. The line with the maximum number of edges is analysed.

Example: Consider the example as seen in Figure 6.2. This is a medical bill which one customer wants to claim for reimbursement against insurance. The total amount of the bill is 2500/- Rupees. But after writing this amount, the same writer, using the same pen (colour: blue), manipulated it and changed the amount to 2509/- Rupees. A similar type of fraud case is also discussed in Vishnu Dev Yadav and Lingan (2023), forgery in the age of the insured person.

6.1.1.3 Type 3: Addition or Alteration of Existing Word Using Same Colour Pen by Different Person

This is the most commonly used forgery technique. Works in Dansena, Bag, and Pal (2017; 2022) explain this type of scenario. Here, the fraudster changes the total authenticity of the cheque by adding or altering it. However, they try to use a similar coloured ink pen with a small amount of pen strokes so that it is hard to find any difference. Investigation mainly

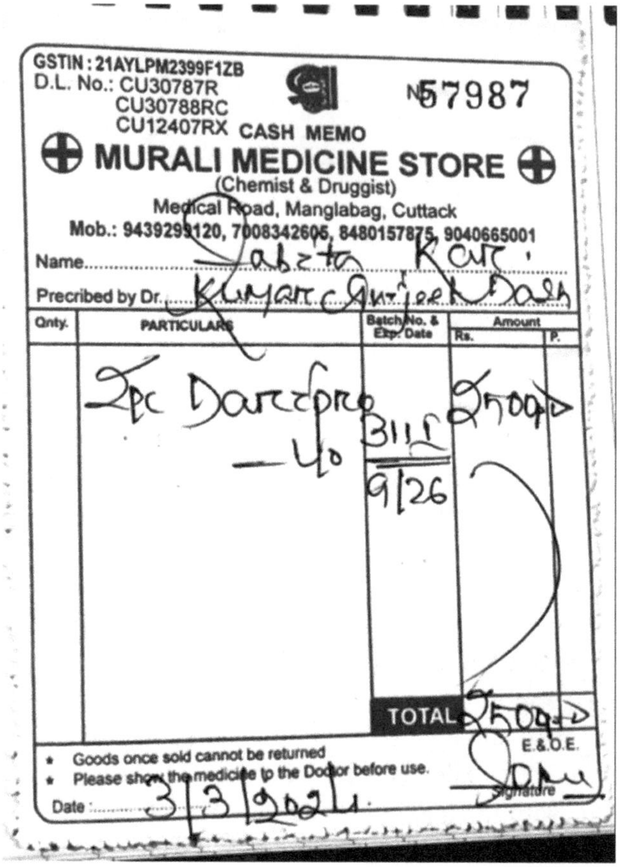

GSTIN : 21AYLPM2399F1ZB
D.L. No.: CU30787R
CU30788RC
CU12407RX CASH MEMO
No 57987

MURALI MEDICINE STORE
(Chemist & Druggist)
Medical Road, Manglabag, Cuttack
Mob.: 9439299120, 7008342606, 8480157875, 9040665001

Name.......
Precribed by Dr.......

Qnty.	PARTICULARS	Batch No. & Exp. Date	Amount Rs.	P.
		B11 9/26		
		TOTAL		

E.&O.E.

- Goods once sold cannot be returned
- Please show the medicine to the Doctor before use.

Date : 3/3/2024

Signature

FIGURE 6.2 Alteration of digit by the same person and same pen.

focusses on the identification of geometrical and structural differences in the features of words. From this, it is easily identifiable whether the writing belongs to the same writer or another, as discussed in Wadhwa, Maheshwari, Dansena, and Bag (2018). Example: In the cheque shown in Figure 6.3, 7 Lakh is added in the amount section before 45,000/- and is marked in a circle. It is very hard to differenciate the original and tampered images.

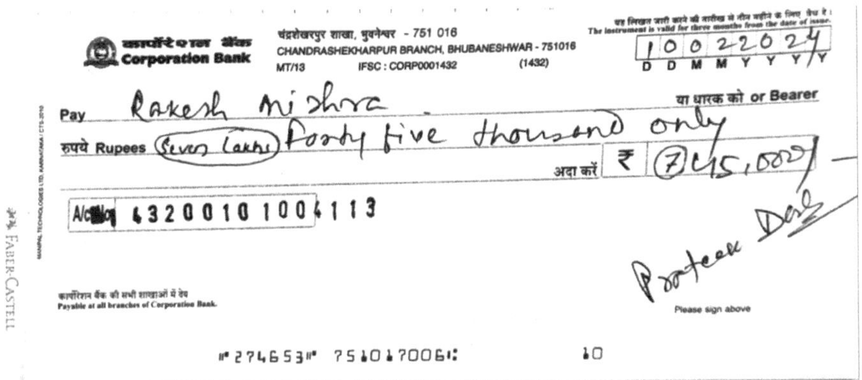

FIGURE 6.3 Addition of new word using the same pen written by a different writer.

6.2 CLASSIFICATION OF EXISTING ANALYTICAL TECHNIQUES

From the perspective of the examiner, there are different scientific methods available to analyse these tiny and cleverly tampered handwritten documents. The analytical technique can be categorised into two categories i.e., either based on the analysis of chemicals used in this method or by passing light of different wavelengths through the written document.

- Based on chemical analysis of the written document technique, it is called chromatography.
- Based on absorption and reflection properties of light, the ink analysis technique is called spectroscopy.
- Analysis of scanned image using image processing techniques is also available.

6.2.1 Chromatography

The most important components of ink are either dyes, which are in solution form called a vehicle, pigment, which is in granular form, or the mixing of both, as explained in Brackett Jr and Bradford (1952). So it's a collection of different chemicals. These chemicals are made up of chemicals and atoms. Chromatography is based on this fact. It will analyse the chemical properties to identify whether the mixture is equal or not. In this method, two phases are considered: one is the mobile phase, which then moves on to the stationary phase to analyse the chemical behaviour of the component, and based on this behaviour, the solution is analysed. Here the stationary phase is set up using various lab methods, while mobile phase is attained by variations in temperature and pressure. It is clearly explained in Allen (2015), Easterby (1993), and Siegel (2013).

6.2.2 Spectroscopy

In Hollas (2004), another technique that has been popularised in forensic science is discussed. It is either based on reflection or absorption properties of light, which is also divided based on different wavelengths when induced through sampled questioned documents. This type of technique is called spectroscopy. As mentioned in Rao (2012), when light is passed through the document, the bonds of the molecules are disturbed, and they behave differently. Either it absorbs the light or it reflects the light. Based on this, the authenticity of the document is analysed. Light of different wavelengths is discussed next:

- Infrared (700 nm–1 mm)
- Raman rays (785 nm)
- X-rays (0.01 nm–10 nm)
- UV rays (100 nm–400 nm)

Absorption of different coloured pens under Raman rays, infrared rays, and X-rays are depicted in Figure 6.4, Figure 6.5, and Figure 6.6, respectively.

6.2.3 Image Processing Techniques

The extraction of features from the scanned image using image processing along with application of existing machine learning approaches can be used for handwriting identification as well as handwritten document analysis. Steps involved in handwritten document analysis-based forgery detection is delineated in Figure 6.7.

As per the methodologies discussed in Dansena, Pal, and Bag (2020) and Samsuryadi and Mohamad (2021), image sources for the handwriting identification technique might come from any digital technology. After being captured by a camera or scanner, the image moves on to the next stage. The series of steps known as pre-processing increases the image's accuracy by

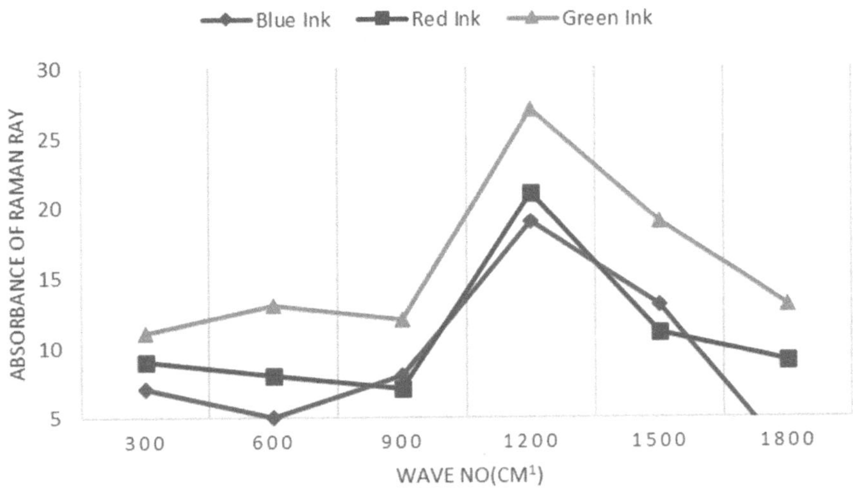

FIGURE 6.4 Raman ray absorption in different coloured pens.

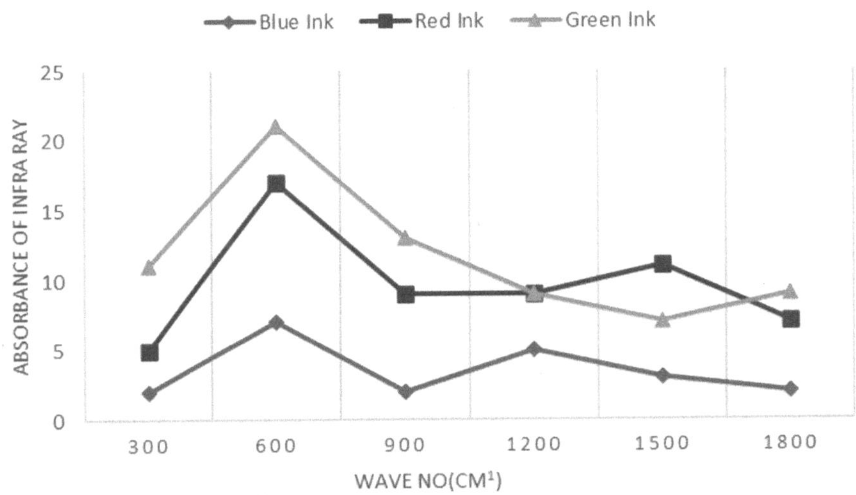

FIGURE 6.5 Infrared ray absorption in different coloured pens.

enhancing its quality. The method of reducing noise from images comes after the handwritten character recognition procedure. This also refers to minimising undesired signals in the image to smooth it out. There exist many algorithms to remove noises in the image. Some of these are the Gaussian filtering method, min-max filtering method, and median filter method. Binarisation is a mechanism of converting greyscale or coloured images to

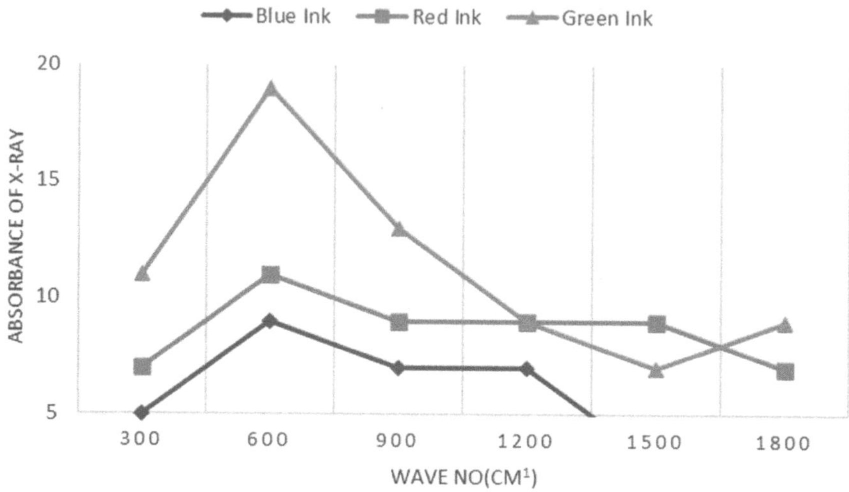

FIGURE 6.6 X-ray absorption in different coloured pens.

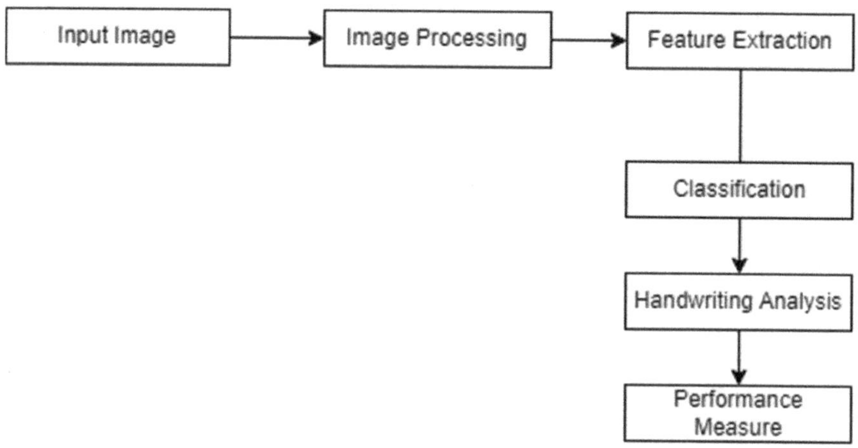

FIGURE 6.7 Steps involved in handwritten document analysis using image processing-based techniques.

binary images. Binary images contain 0 or 1 at each of the spacial locations i.e., pixels. Based on a fixed value, the pixels in images are divided into 0s and 1s. The value of the pixel is replaced with 0 or with 1 if it is smaller than the constant. The process of enlarging or contracting an image is known as morphological operation. The algorithm would anticipate a consistent image size, which is the primary reason this is done. An image can have its

TABLE 6.2 Classification of chromatography techniques

Type	Techniques	Classification
1	Chromatography	•Thin layer chromatography •High performance liquid chromatography •Gas chromatography
2	Spectroscopy	•Infrared spectroscopy •Raman spectroscopy •X-ray spectroscopy
3	Image processing of scanned documents	•Mahalanobis distanced •HSV colour space •Texture analysis •Support vector machine •Ink feature •Neural network •MLP classifier •*k*-NN

border filled with pixels to make it larger. We can eliminate pixels from an image's border to make it smaller. The algorithm for recognition begins with feature extraction, which is a crucial step in the process. Every character has distinctive qualities. It consists of a set of rules, each of which describes a particular quality of a character. This phase involves the extraction of such traits. Once categorisation is finished, training is finished and input data testing begins. The testing data would pass each step of the procedure described above, and the matching rules would be allocated different probabilities. Once the rule with the highest probability has been determined, the associated class-label is recognised as a character.

6.3 DETAILED DISCUSSION OF EXISTING TECHNIQUES

This section delineates each of the above-mentioned techniques in detail with respect to each other. Moreover, further classification of each technique is discussed in Table 6.2.

6.3.1 Types of Chromatography-Based Handwritten Document Analysis Techniques

6.3.1.1 Thin Layer Chromatography (TLC)

TLC is based on the separation principle, where separation is achieved through attraction to attain stability among different phases mentioned. As already described, here two phases are considered: one is mobile and the other is stationary. In this process, a small amount of ink is diluted in a solvent. Then, by taking a small sample, it is placed on a TLC plate, which is a layer of silica gel on a glass plate. Then this TLC plate is placed into a jar containing ethanol, which is considered a mobile phase. After some time, based on the capillary rise principle, the solvent will rise and separate into different colours. From this, the examiner can separate it and analyse it. This will give the RF (retention factor) value. Detailed calculations of the RF value are mentioned in the literature. This will give quantitative data, which is not so accurate. The major drawback is that it may not give correct results in some cases.

6.3.1.2 High Performance Liquid Chromatography (HPLC)

This process is a small variation of the TLC procedure. It is more efficient as it provides rapid chromatograms with high resolution and requires a small sample amount. As noted in El-sabbah et al. (2019), Kher, Mulholland, Green, and Reedy (2006), and Renger, Végh, and Ferenczi-Fodor (2011), by following this procedure, a thin glass sheet of the TLC plate is replaced with a column in a tubular shape filled with stationary phase material. The solution containing the sample ink mixed with the mobile phase is passed through it with the help of a high-power pump. Here, the pressure and temperature are variable. Then at the other end, the stream of solvent is collected and analysed using various wavelengths of light. It also gives quantitative results.

6.3.1.3 Gas Chromatography (GC)

This is a slightly varied technique. As discussed in Alpert, Keiser, and Szymanski (2012), El-sabbah et al. (2019), and Hu, Zhang, and Diao (2023), the sample mixed with a solvent is converted into gas at high temperatures. Then it is mixed with hydrogen gas, helium gas, nitrogen, or air to

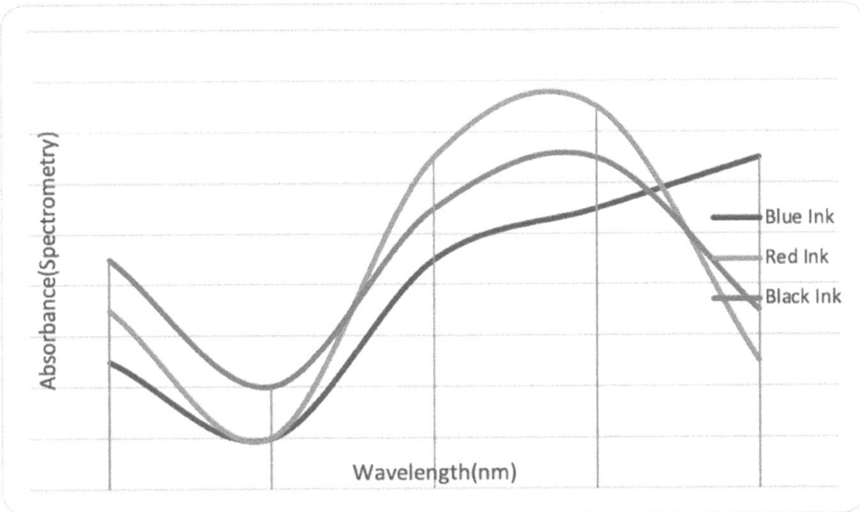

FIGURE 6.8 Ink identification by chromatography technique based on wavelength.

represent a mobile phase. It is transferred through the tube coated with stationary phase material like silica. Then, according to chemical reaction and affinity to the stationary phase, it is separated, and based on that, the components of inks are analysed. Here the K_c value, which represents distribution constant, is calculated. The K_c value is affected by the temperature and the nature of the stationary phase substance. It is a highly accurate quantitative analysis. The major drawback it faces is that the solution degrades at high temperatures. It always works with mass spectroscopy (MS). In GS, when the mass to charge ratio of the molecules is calculated, it is termed MS.

Details of the chromatography and spectroscopy techniques considering different colour pen are presented in Figure 6.8 and Figure 6.9.

6.3.2 Types of Spectroscopy-Based Handwritten Document Analysis Techniques

6.3.2.1 Infrared Spectroscopy

This is a non-destructive type of analysis. The wavelength of infrared normally ranges from 700 nm to 1 mm. Its wavelength is a bit longer than visible light. Based on emission, absorption, and reflection of molecules, it is analysed as mentioned in Alpert, Keiser, and Szymanski (2012) and Kudelski

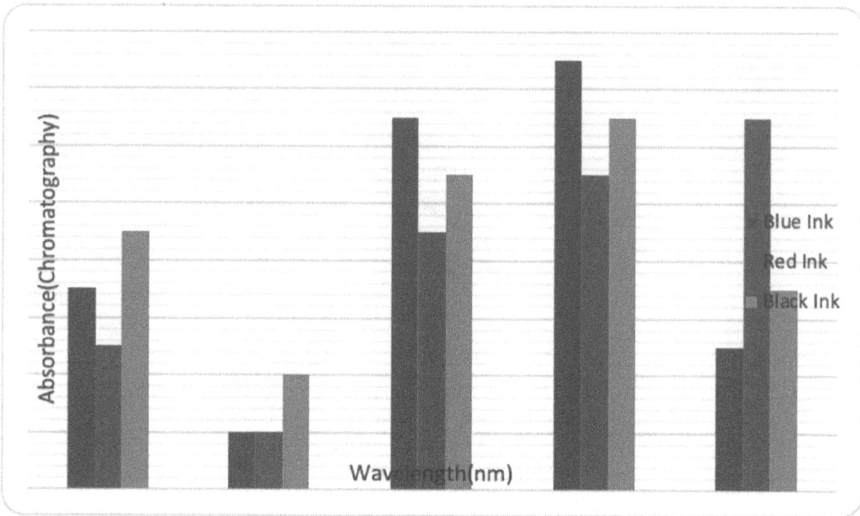

FIGURE 6.9 Ink identification by spectroscopy technique based on wavelength.

(2008). When the sample paper is shown to infrared light, the natural bond of molecules is disturbed, and it vibrates either symmetrically or asymmetrically. It also absorbs and reflects some light, and by analysing that, the examiner can conclude the test. When the frequency is the same, it absorbs the light.

6.3.2.2 Raman Spectroscopy

Based on research performed by Das and Agrawal (2011), infrared spectroscopy is not feasible in some cases; for instance, the presence of water cannot affect the quality of the collected signal, and it also provides some additional information. In this technique, light is absorbed when the frequency of the incident light is equal to the energy level gap between two energy levels. It gets scattered when the photon interacts with molecules. It faces problems with fluorescence caused by laser light.

6.3.2.3 X-Ray Spectroscopy

X-ray has a shorter wavelength with great energy. When this high-energy incident light reaches a substance, it gets ionised by losing electrons, according to De Groot (2005) and Zimmermann et al. (2020). This space of the lost electrons is filled by other high-power electrons, which also emit light. In this case, the examiner concludes by comparing the wavelength of emitted light with the wavelength of emitted light from a known substance. The known substance's emitted wavelength is already stored in the database. This is a very efficient form of analysis because there are variations of X-rays present in the market. It can be applied to pigment and element analysis for provenance and authentication investigations in artworks. Archaeological studies can make use of it. Additionally, it can help with quality control in the printing sector. X-ray fluorescence is prized for its quick, semi-quantitative to quantitative elemental analysis as well as its non-destructiveness.

6.3.3 Classification Using Image Processing-Based Techniques

6.3.3.1 Classification Using Mahalanobis Distanced-Based Clustering

The concept of this classification is Mahalanobis distance, which was first proposed by P.C. Mahalanobis. One digital scanned image is represented as a basic unit that is pixel. Further, pixels are classified or converted to clusters by using the Euclidean method to detect the fraudulent case (da Silva Barbosa, Lins, De Lira, & Camara, 2014). It normally gives the distance between the two samples of pen by calculating the summation of pixel-wise intensity level. One sample is the original pen writing and another is altered pen writing. But a problem will arise in the case of overlapping of handwriting, because in this case the Euclidean value is very small. To overcome this, a new method is used, the Mahalanobis method. In this, it measures the distance between a point and the distribution. Here the distribution is considered because the intensity of ink throughout the handwriting is not same. Here, red, green, and blue (RGB) pixels are considered. This process includes the following:

(1) First calculate the RGB component of each individual vector.

(2) Segregate the pixels to two populations.

(3) Calculate the co-variance for each population.

(4) Compute the inverse pooled sample covariance matrix using formula.

(5) Vector of difference is calculated between the two populations.

(6) Then by following the formula, calculate the Mahalanobis distance.

It is also considered in the case of analysing aged ink to check for forgeries based on the principle that paper normally gets oxidised and vaporised when exposed to light even if it is placed in an unventilated room. But handwriting from the same person is hard to differentiate through this technique.

6.3.3.2 Classification Using HSV Colour Space

This HSV (hue, saturation, value) model is available in high-end image editing software. It represents the colour and intensity of the ink, which depends upon various factors like absorbing capacity of paper, angle of pen, and properties of ink and paper. In this method, a document containing a sizable amount of words is scanned through a high-end scanner. Then it is divided into pixels. It is converted to HSV by following an equation. From this, a histogram is followed by taking threshold value into consideration so that background and foreground pixels get separated. As experimented by Chakravarthy and Haritha (2005) and Dasari and Bhagvati (2007), the examiner has to find the saturation value from the position having highest saturation, and *h* which is the highest number of pixels having high saturation. Using this data, a histogram of saturation has to be calculated as Gaussian distribution. This distribution will give an F-ratio value, which is the deciding factor for classification.

6.3.3.3 Classification by Texture Analysis and Histogram Matching

In local binary pattern (LBP), Gabor filters are used to construct the histogram (Gorai, Pal, & Gupta, 2016). In LBP, contrast is measured along with the relationship with neighbouring pixels. In this method, a sliding window of square pixel units is used. It continuously produces 8-bit binary digits, which are converted to decimal value. From this produced LBP value, a histogram is created and features are analysed. By using the GB method, the image can be easily analysed by passing it through the process, which results in a complex sinusoidal wave with Gaussian envelope. A histogram is constructed from the obtained features and becomes normalised. Matching score (MS) is used to calculate the similarity between two features. Its value may be 1 or 0. Then, by taking the average of this MS value, it is compared with one confirmed threshold value, which allows a conclusion to be reached as to whether different or the same ink had been used.

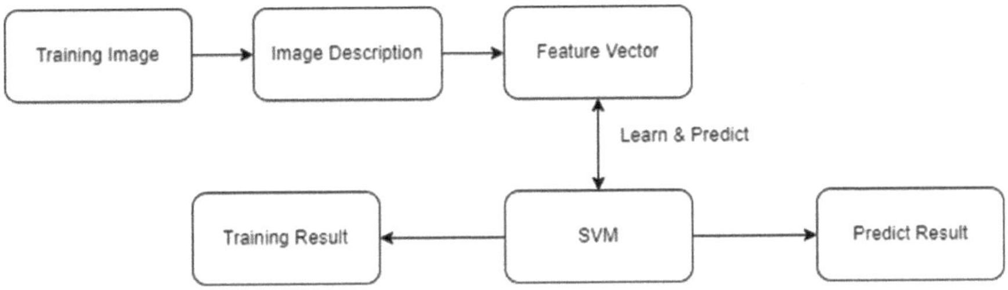

FIGURE 6.10 SVM classification model for handwritten document analysis.

6.3.3.4 Classification of Colour Values by Support Vector Machine (SVM)

Through this type of technique, an examiner can determine whether the given sample is from the same pen or a similar coloured pen (Berger, 2013). The procedure to determine this is divided into two parts. First, by analysing the feature vector of ink, the examiner can calculate quantitative amount. Then, by analysing the image, the examiner obtains a qualitative discrimination of the image. This method is called deconvolution. Elaborating this procedure, samples are first collected. Then the steps of feature extraction are followed by analysing the feature vector. Eucledian distance is calculated between every possible pair of handwriting samples, the histogram is drawn, and the likelihood ratio is calculated by using simple deviation. The image processing technique then follows. In this method, we use a 2D histogram. Then, by using a support vector machine (SVM), the examiner can classify the difference in samples, as shown in Figure 6.10.

6.3.3.5 Classification by Analysing Ink Feature

The features of the colour are based on the chemical and physical properties of the ink (Berger, 2013). Every ink has its own properties, and by analysing the typical features, the examiner can classify the ink. After this process, all the images are collected, scanned, and converted to binary. The connected region is specified according to this. Then features are calculated based on RBG intensity, standard deviation of RGB, and skewness of RGB. Distance is measured for each feature using the RMSE method. Outlier points are detected using the statistical method of the Tomson Tau test. Finally, by using anomaly analysis, the examiner can classify the tampered handwriting.

6.3.3.6 Classification of Non-Black Pen Using HSV Colour Space

In Kumar, Pal, Sharma, and Chanda (2009), the researchers proposed two class pattern recognition techniques using image processing for solving document forensic problems. This process follows feature extraction and alteration detection. The *k*-nearest neighbour classifier is used for alteration detection. In this approach, the authors used an MLP-based feature selection method.

Feature extraction: To compute different characteristics, the opponent chromaticity space (rg, yb) and selected YC_bC_r were used. The opponent process theory of colour vision provides a rationale for choosing two colour spaces at once. Not all features of colour vision are explained by the trichromatic theory of colour vision, according to that theory. Yellowish-blue opponent neurons regulate the opponent responses, which is why combinations like reddish-green or yellowish-blue, though they can exist in YC_bC_r space, are not seen.

Alteration detection: According to its durability and simplicity, the nearest neighbour classifier is employed to detect modifications. The training data is derived from the data for the remaining nine pens, with one specific pen P_k being excluded. For every pen P_k ($k = 1, \cdots, 10$), the procedure is repeated. One can prevent bias against any pen by doing this.

Feature analysis: Use a multilayered perceptron (MLP)-based feature selection method. Here each input node modulates the input feature value by computing the product of the input feature and the gate function value. The learning process uses gradient descent to minimise the classification error. The advantages of this method over many others are that it can account for the non-linear interaction between the features as well as that between features and the tool neural network.

6.3.3.7 Classification Based on Ink Stroke Using SVM and Neural Network

In a method proposed in Kumar, Pal, Chanda, and Sharma (2009), a neural network-based feature analysis and SVM-based detection technique for identifying the alteration in ballpoint pen strokes were used. The proposed approach improves the performance of the system.

Feature extraction: The next phase extracts some traits that can be utilised to differentiate between different pens and are similar to what humans can perceive. We have selected the YC_bC_r colour space with this in mind. Colour and texture features from each channel are the features we plan to extract for the current issue. For colour features and texture characteristics, we have calculated different moments of joint distribution of different colour components and several aspects of the grey-level co-occurrence matrix (GLCMS).

Feature analysis: When a system is identified with ease and produces superior generalisation, it is generally because a set of just sufficient features is being used. Learning and decision-making times are typically shortened when a narrow collection of features is required. While simultaneously optimising the main goal, the goal is to identify the smallest possible subset of features. To reduce the detection error, the learning process uses gradient descent. As a result, the gates linked to excellent features that can decrease error more quickly are probably going to open more quickly, whereas the gates linked to bad features—those that are unable to reduce error—are probably going to close more tightly.

Support vector machine-based solution: This algorithm for supervised learning has two classes. To resolve the optimisation issue, SVM determines the ideal weight vector and bias. The training vectors are implicitly transferred to a higher dimensional space using a suitable kernel function when the dataset is not linearly separable, and the separating hyperplane is then found in that high dimensional space. SVM-based feature analysis is expected to enhance the model's performance.

6.3.3.8 Classification of Ink According to Ballpoint Pen Stroke Using k-NN, SVM, and MLP

A comparative study of different classifiers i.e., *k*-NN, MLP, and SVM, to treat detection of the alteration in pattern recognition problems was proposed in Kumar, Pal, Chanda, and Sharma (2011). From the analysis, it was found that SVM was the most successful one. The proposed approach is useful in forensic science in combination with conventional techniques.

Feature extraction: In most cases, it is exceedingly challenging to distinguish between distinct strokes, even after the pictures have been enlarged. Nonetheless, there could be some colour differences between two blue inks because different producers might employ various chemical mixtures to make the ink. The feature set must take on two duties in order to handle such challenges. Initially, it ought to depict the subjective distinctions amongst the inks. Secondly, the device must to have the ability to identify variations

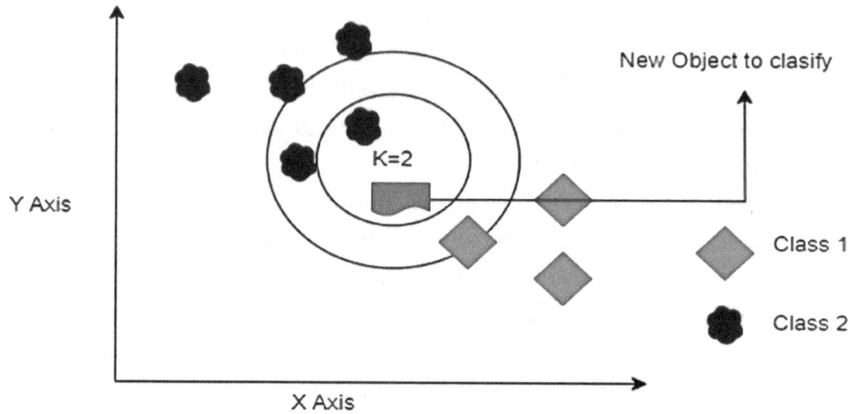

FIGURE 6.11 *k*-NN-based classification.

in texture that can emerge from the interaction between the pen and paper. We have decided to use the YC_bC_r colour model to fully describe colour.

Feature analysis: Minimal feature analysis is used to maximise the recognition score or forecast accuracy by identifying a subset of features. A modest set of features improves system comprehension and aids in locating elements important to the goal. Therefore, in this case, feature analysis is required.

k-NN classifiers: A straightforward classifier, the *k*-nearest neighbour classifier, is employed by the authors to attempt to identify the modification. Using all 106 features for various selections of features, they finally execute the *k*-NN algorithm. While there is a discernible range in accuracies among the individual pens, the average accuracy of all six *k*-NN classifiers is practically identical. These modifications are for subsequent discussion.

SVM and MLP classifiers: The graph was plotted by the author to compare the classifier. Using a neural network-based feature selection method that captures the subtle nonlinear interaction between features, the classifiers' performance is improved. Based on a comparative review of the classifiers' performances, all three of the models under study produce pretty excellent results; SVM is the most effective model among the three. In addition to traditional procedures, the forensic community can greatly benefit from such a non-destructive technique. A pictorial representation of *k*-NN and MLP is depicted in Figure 6.11 and Figure 6.12.

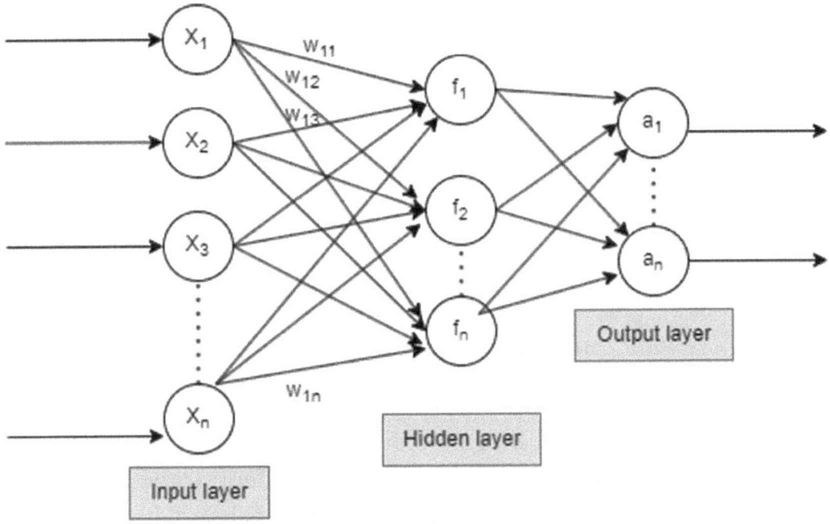

FIGURE 6.12 MLP-based classification.

6.3.3.9 *Ink Classification Using k-NN*

Mendhe and Ratnaparkhe (2018) proposed a radiation-based recognition technique for ballpoint pen strokes. They used the GLCM-based feature extraction, *k*-NN, and SVM-based classification methods. *k*-NN produced better identification.

Pre-processing: We reduced our database photographs during pre-processing, and we employed a median filter to eliminate input image noise. To extract the features from the input database photos, the image's contrast is then changed, and its RGB (or colour) to greyscale conversion is made.

Feature extraction: A key factor in pattern recognition techniques is the extraction of an image's characteristics. Even with image enlargement, it is frequently impossible to distinguish between pens made by different companies using the same ink. Considering that different manufacturers prepare ink using different chemical compositions, there will always be some colour variance.

k-NN classifier: The higher number of votes from neighbouring pixels determines the identification or classification of a case. Based on the most common feature among the *k*-nearest neighbours determined by the distance function, the class is allocated to a case. Euclidean distance is the distance function that is typically used to determine the k-nearest neighbour.

SVM classifier: By creating the most feasible hyperplane between the two classes, this classifier separates all of the data points of one class from the other. The greatest distance between the support vectors of the two classes is used to determine which hyperplane is most appropriate. Whichever data points from both classes are closest to the hyperplane are the support vectors.

6.3.3.10 Classification Using MLP Classifier

In Roy and Bag (2019), an MLP-based data instance classification was proposed by extracting colour-based statistical features. This approach is useful for handwritten document analysis as well as ink identification.

Image acquisition: Insufficient target scanning is a problem that image enhancement cannot fix. Technology has advanced to the point that this process is now simpler thanks to devices like PDAs, digital cameras, and scanners. The intended goals might not be achieved if an object is not acquired correctly. As a result, a document laser scanner is used to take pictures at this location.

Image pre-processing: Two primary procedures are involved—word-based segmentation (i) involves manually breaking each word in the scanned document into smaller images of a single word, and (ii) eliminates background to reduce the effect of background.

Foreground extraction: The k-means binarisation algorithm has been implemented to extract only the ink pixels. The binarised image is masked to extract only the object pixels from each colour plane.

Feature extraction: A small number of statistical features are extracted at this stage from each and every processed image. To compute features, the YC_bC_r colour model is selected. Every colour plane has its features removed.

Feature classification: We select the multilayer perceptron classifier (MLP). Compared to other classifiers, MLP aids in sorting, and yields more accurate results. Even in cases where features interact nonlinearly, the MLP classifier is able to make decisions that are more accurate.

6.3.3.11 Classification of Ink Using k-Means Binarisation

Roy and Bag (2022) proposed a non destructive approach to differentiate similar ink of different brands. This technique is useful in identifying fraudulent alteration of handwritten forensic documents. The authors used the MLP classifier and compared the performance of both known and unknown datasets.

k-Means binarisation: Choose at random the centre points of each partition and set *k*'s initial value to 2. *k* must be set to two to divide the image's pixel count into two labels: one label for the object pixels and another for the background pixels. Create a new cluster by placing the pixel's greyscale intensity in the closest centre. The Euclidean distances between the pixel values and the cluster centres are used to choose which closest cluster to assign the pixel values to. Again, use the allocated pixel intensity mean to figure out the clusters' new centres. Repeat step 2 and step 3 until the centre values remain unchanged. Two centres are averaged to determine the threshold value.

Feature extraction: A feature set can help overcome obstacles of this nature. Furthermore, the process of identifying ink distinction is impacted by the employment of colour models. In this investigation, the YC_bC_r colour model is used. Each word's image is separated into fixed-size cells.

Feature classification: Multiple pens with different individual but similar-coloured inks have been identified using an MLP classifier. Using a training dataset, the MLP classifier was trained. The computational neural network is trained using feature vectors of various word pairings to get the desired level of accuracy for a predetermined dataset. The same MLP model is used to test the unknown dataset using validation set testing once the training phase is complete.

6.3.3.12 Classification Using CNN

A method has been proposed in Dansena, Pramanik, Bag, and Pal (2019) that efficiently detects forms of alteration in existing handwriting. This method utilises transfer learning-based CNN. Further, another approach has been proposed to generate synthetic word alteration dataset to train large CNN models (Dansena, Bag, & Pal, 2021). The fraudulent alteration of handwriting can also be detected using a CNN technique without any analysis of pen stroke and the properties of paper and ink. This technique was described as a landmark in image processing techniques. In this technique, two types of datasets are used. These are either pre-trained by ImageNet or trained from scratch. Alexnet, VGG-16, VGG-19, and MobileNet are the major algorithms followed in this method. But the main drawback of this method is that it requires a large input data which might not always be possible. To solve this difficulty, it is suggested that a synthetic dataset is created by comparing spacing, slant, geometric formation, and interpolation in collected sample handwriting.

In this methodology, total image is converted 227* 227*3 RGB pixels. Then it is applied to the first convolutional layer with stride of 4. After that, the maxpooling operation is performed. The output of first layer is again treated as input in second convolutional layer. At the final stage, the fully connected layer is concluded followed by a SoftMax layer. To counter linearity in input image activation, functions like ReLu and weights are used. Alexnet comprises of five convolutional layers, VGG-16 contains 13 convolutional layers, VGG-19 contains 16 convolutional layers, and MobileNet contains 28 convolutional layers, followed by three fully connected layers.

Even though there are a number of models available for identification or alteration of the original document, both non-destructive and destructive methods require a physical copy of the documents. Non-destructive techniques include a spectroscopy method that can only scan the physical documents. However, in the case of insurance claims, medical bill claims, and many other cases, the scanned documents are always required to be uploaded. It's very difficult for the examiner to verify the authentication of these scanned documents. In that case, the examiner can effectively use the image processing technique.

6.4 CONCLUSION

Document forensics plays a crucial role in legal investigations and fraud prevention by employing meticulous analysis techniques to ascertain the authenticity and characteristics of various documents. The expertise of qualified document examiners is instrumental in providing reliable opinions on handwriting, signatures, ink and paper composition, printing processes, alterations, and typewritten documents. Through these analyses, document forensics contributes to upholding the integrity of legal proceedings and ensuring the accuracy of information presented in various contexts. As technology and methods in this field continue to advance, document forensics remains an essential tool in uncovering the truth and maintaining the credibility of written evidence. This chapter discussed existing methods for handwritten document analysis. Initially, the different techniques used by the forger were discussed to highlight the issues in handwritten document analysis. Further, existing methods were classified into two categories—light-based and chemical analysis—to highlight the materialistic analysis associated with each of the methods. Further, chromatography-based methods were discussed. Thereafter, spectroscopy-based existing methods were discussed. It is important to note that chromatography-based techniques are chemical-based materialistic analyses that often damage the physical document itself. Moreover, spectroscopy-based methods require specific hardware for analysis purposes, which is often costly and rarely available to every organisation dealing with handwritten documents. Hence, existing techniques in handwritten document analysis utilising normal scanning devices and machine learning-based methods were further discussed in detail highlighting the advantages over other existing methods.

REFERENCES

1. Adam, C. (2016). *Forensic evidence in court: Evaluation and scientific opinion*. John Wiley & Sons.
2. Allen, M. J. (2015). *Foundations of forensic document analysis: Theory and practice*. John Wiley & Sons.
3. Alpert, N. L., Keiser, W. E., & Szymanski, H. A. (2012). *IR: Theory and practice of infrared spectroscopy*. Springer Science & Business Media.
4. Berger, C. E. (2013). Objective ink color comparison through image processing and machine learning. *Science & Justice*, 53(1), 55–59.
5. Brackett Jr, J. W., & Bradford, L. W. (1952). Comparison of ink writing on documents by means of paper chromatography. *Journal of Criminal Law, Criminology and Police Science*, 43, 530.
6. Chakravarthy, B., & Haritha, D. (2005). Classification of liquid and viscous inks using HSV color space. In *Proceedings of 8th International Conference on Document Analysis and Recognition* (pp. 660–664). IEEE. da Silva Barbosa, R., Lins, R. D., De Lira, E. D. F., & Camara, A. C. A. (2014). Later added strokes or text-fraud detection in documents written with ballpoint pens. In *2014 14th International*

Conference on Frontiers in Handwriting Recognition (pp. 517–522). IEEE.

7. Dansena, P., Bag, S., & Pal, R. (2017). Differentiating pen inks in handwritten bank cheques using multi-layer perceptron. In *International Conference on Pattern Recognition and Machine Intelligence* (pp. 655–663). Springer.
8. Dansena, P., Bag, S., & Pal, R. (2021). Generation of synthetic data for handwritten word alteration detection. *IEEE Access*, 9, 38979–38990.
9. Dansena, P., Bag, S., & Pal, R. (2022). Pen ink discrimination in handwritten documents using statistical and motif texture analysis: A classification based approach. *Multimedia Tools and Applications*, 81(21), 30881–30909.
10. Dansena, P., Kumar, K. P., & Pal, R. (2015). Line based extraction of important regions from a cheque image. In *2015 Eighth International Conference on Contemporary Computing (IC3)* (pp. 183–189). IEEE.
11. Dansena, P., Pal, R., & Bag, S. (2020). Quantitative assessment of capabilities of colour models for pen ink discrimination in handwritten documents. *IET Image Processing*, 14(8), 1594–1604.
12. Dansena, P., Pramanik, R., Bag, S., & Pal, R. (2019). Ink analysis using CNN-based transfer learning to detect alteration in handwritten words. In *International Conference on Computer Vision and Image Processing* (pp. 223–232). Springer.
13. Das, R. S., & Agrawal, Y. (2011). Raman spectroscopy: Recent advancements, techniques and applications. *Vibrational Spectroscopy*, 57(2), 163–176.
14. Dasari, H., & Bhagvati, C. (2007). Identification of non-black inks using HSV colour space. In *Ninth International Conference on Document Analysis and Recognition (ICDAR 2007)* (Vol. 1, pp. 486–490). IEEE.
15. De Groot, F. (2005). Multiplet effects in X-ray spectroscopy. *Coordination Chemistry Reviews*, 249(1–2), 31–63.
16. Easterby, D. (1993). Analysis of printing inks. In *The printing ink manual* (pp. 865–900). Springer. El-sabbah, M. M., Gomaa, A. Z., Al-Hawary, A. S., et al. (2019). Dating the ballpoint pen inks using gas chromatography-mass spectrometry technique. *Egyptian Journal of Chemistry*, 62(3), 385–400.
17. Gorai, A., Pal, R., & Gupta, P. (2016). Document fraud detection by ink analysis using texture features and histogram matching. In

2016 International Joint Conference on Neural Networks (IJCNN) (pp. 4512–4517). IEEE.
18. Hollas, J. M. (2004). *Modern spectroscopy*. John Wiley & Sons.
19. Hu, X., Zhang, X., & Diao, Z. (2023). Rapid and simultaneous determination of nine volatile solvents in ink entries with improved gas chromatography. *Microchemical Journal*, 190, 108637.
20. Kher, A., Mulholland, M., Green, E., & Reedy, B. (2006). Forensic classification of ballpoint pen inks using high performance liquid chromatography and infrared spectroscopy with principal components analysis and linear discriminant analysis. *Vibrational Spectroscopy*, 40(2), 270–277.
21. Koppenhaver, K. M. (2007). *Forensic document examination: Principles and practice*. Humana Press.
22. Kudelski, A. (2008). Analytical applications of raman spectroscopy. *Talanta*, 76(1), 1–8.
23. Kumar, R., Pal, N. R., Chanda, B., & Sharma, J. D. (2009). Detection of fraudulent alterations in ball-point pen strokes using support vector machines. In *2009 Annual IEEE India Conference* (pp. 1–4).
24. Kumar, R., Pal, N. R., Chanda, B., & Sharma, J. D. (2011). Forensic detection of fraudulent alteration in ball-point pen strokes. *IEEE Transactions on Information Forensics and Security*, 7(2), 809–820.
25. Kumar, R., Pal, N. R., Sharma, J., & Chanda, B. (2009). A novel approach for detection of alteration in ball pen writings. In *Pattern Recognition and Machine Intelligence: Third International Conference, PReMI 2009 New Delhi, India, December 16–20, 2009 proceedings 3* (pp. 400–405). Springer.
26. Mendhe, S. S., & Ratnaparkhe, V. R. (2018). Fraudulence detection using image processing approach. In *2018 2nd International Conference on Intelligent Computing and Control Systems (ICICCS)* (pp. 1539–1542). IEEE.
27. Rao, K. N. (2012). *Molecular spectroscopy: Modern research*. Elsevier.
28. Renger, B., Végh, Z., & Ferenczi-Fodor, K. (2011). Validation of thin layer and high performance thin layer chromatographic methods. *Journal of Chromatography A*, *1218*(19), 2712–2721.
29. Roy, P., & Bag, S. (2019). Forensic performance on handwriting to identify forgery owing to word alteration. In *2019 IEEE 5th International Conference on Identity, Security, and Behavior Analysis (ISBA)* (pp. 1–9). IEEE.

30. Roy, P., & Bag, S. (2022). Ink analysis based forensic investigation of handwritten legal documents. *Multimedia Tools and Applications*, 81(16), 23007–23047.
31. Samsuryadi, R. K., & Mohamad, F. S. (2021). Automated handwriting analysis based on pattern recognition: A survey. *Indonesian Journal of Electrical Engineering and Computer Science*, 22(1), 196–206.
32. Siegel, J. (2013). *Ink analysis*. Academic Press.
33. Ubelaker, D. H. (2019). New directions in forensic. *The Future of Forensic Science*, 1, 1. John Wiley & Sons.
34. Vishnu Dev Yadav, K. P., & Lingan, L. (2023). Bank cheque fraud and cheque truncation system to illuminate the fraudulent transaction: A case study. *Journal of Forensic Research*, 14(4), 1–4.
35. Wadhwa, A., Maheshwari, M., Dansena, P., & Bag, S. (2018). Geometrical and structural features for forensics in handwritten bank cheques. In *2018 15th IEEE India Council International Conference (INDICON)* (pp. 1–6). IEEE.
36. Zimmermann, P., Peredkov, S., Abdala, P. M., DeBeer, S., Tromp, M., Müller, C., & van Bokhoven, J. A. (2020). Modern X-ray spectroscopy: XAS and XES in the laboratory. *Coordination Chemistry Reviews*, 423, 213466.

Index

E

F

G

H

I

K

L

M

FSC
www.fsc.org
MIX
Papier | Fördert gute Waldnutzung
FSC® C083411

Kolibri 360 GmbH
Ferdinand-Jühlke-Straße 7
99095 Erfurt, Deutschland
produktsicherheit@kolibri360.de